MCQ TUTOR
for students of
MICROBIOLOGY

MCQ TUTOR for students of MICROBIOLOGY

John Gordon B.A., (Cantab), M.B., Ch.B., (Edin.)

Consultant Microbiologist, Gartnavel General Hospital, Glasgow, and Honary Clinical Lecturer, University of Glasgow

William Heinemann Medical Books Ltd
London

First published 1978
Reprinted 1985

ISBN 0 433 12405 9

Text set in 10/11 pt IBM Century, printed by photolithography,
and bound in Great Britain by Antony Rowe Ltd, Chippenham

CONTENTS

Preface vii
Introduction ix

Part One
MICROBIAL BIOLOGY AND HOST DEFENCES

1. Historical 1
2. Bacterial Cytology 3
3. Bacterial Physiology and Genetics 9
4. Host-Parasite Relationships 21
5. Mechanisms of Pathogenicity 27
6. Host-Defence Mechanisms 35
7. Characteristics of Viruses 55
8. Disinfection and Sterilization 61

Part Two
MICRO-ORGANISMS OF MEDICAL IMPORTANCE

1. Pyogenic Cocci 67
2. Gram-Positive Bacilli 79
3. Mycobacteria, Nocardia and Actinomyces 87
4. Coliform Bacilli 95
5. Small Gram-Negative Bacilli 109
6. Spirochaetes 116
7. Rickettsiae, Chlamydiae and Mycoplasmas 119
8. Fungi 125
9. Viruses 129

Part Three
MICROBIAL DISEASES AND CLINICAL MICROBIOLOGY

1. Transmission of Infective Disease 133
2. Bacterial and Mycotic Infections 147
3. Virus Infections 185
4. Laboratory Diagnosis of Infection 203
5. Antimicrobial Agents 225
6. Prophylactic Immunization 239

Part Four
OTHER FORMS OF MULTIPLE CHOICE QUESTIONS 247

PREFACE

Multiple choice questions are now generally accepted as a valid and reliable method for objective assessment of knowledge and understanding after a period of instruction. Objective examinations are well suited to testing factual knowledge in scientific and technical subjects and have been widely applied as a supplement to the traditional essay paper in university, college and registration board examinations in microbiology.

This book comprises some 400 MCQ's with annotated answers, mainly in bacteriology but also including immunology, virology and parasitology, with the emphasis placed on clinical applications. It is intended primarily for medical students in the second and third undergraduate years as an aid to preparation for the third MB examination, but it should also prove useful to students of medical and laboratory sciences as well as candidates for diploma and college examinations. The book is designed not only to provide a ready means of self-assessment but also to give instruction in tutorial form, hopefully furnishing a starting point for further studies and discussion. It should be used in conjunction with lecture notes and a good textbook to identify particular areas of weakness while proceeding through a course of instruction.

The contents closely follow the conventional arrangement of most courses and have been divided into four parts for ease of reference. Part One is essentially introductory, covering the preclinical aspects of microbiology and host defence mechanisms. Part Two deals with microbial systematics; detailed properties of the various micro-organisms have been largely restricted to those active in the infection process or important in identification, in line with the current trend towards more clinically orientated courses. Part Three has a strong clinical bias and is devoted to the nature and manifestation of a wide spectrum of infective diseases, plus their laboratory diagnosis, treatment and prevention. Part Four is a modest single chapter section which presents examples of a few of the other forms of MCQ that might be met.

Acknowledgements

There remains the pleasant duty of acknowledging my indebtedness and thanks to those of my colleagues who have generously assisted me in the preparation of this small volume. I wish particularly to express my warmest gratitude to Professor Bernard Lennox, who laid a firm foundation to this series in his companion volume MCQ Tutor for Students of Pathology, for much helpful advice and encouragement. I also gratefully acknowledge the help given by other friends and colleagues, especially Dr D. A. R. Simmons, Mr B. A. S. Wilson, Dr T. H. Pennington and Dr. I. C. McKay, whose comments and constructive criticism were invaluable. Finally, my thanks are due to my wife for her continual encouragement.

Gartnavel General Hospital.
Glasgow.

John Gordon.

INTRODUCTION

Examinations serve as standards of reference against which achievement and proficiency in a subject may be measured and in view of the important role they are called on to serve they must utilize methods which satisfy the requirements of validity, reliability and objectivity.

The traditional written examination is the essay paper, but whilst it is acknowledged as the best means of testing ability to marshal facts, construct a reasoned argument and express thoughts clearly it is defective in reliability and objectivity due to lack of an absolute and impartial method of correction. This most serious limitation to the usefulness of the essay paper has been largely removed by the introduction of the combined essay/ MCQ written examination since the MCQ paper affords a satisfactorily valid and reliable objective measure of factual knowledge which frees the essay paper to assess not merely technical accuracy but also the range and quality of ideas in a well presented written narrative. Moreover, the objective paper can more closely reflect the entire range of the course syllabus than can the essay paper which by its very nature is limited in coverage and time consuming.

Objective examinations.

MCQ's have been set in an ingenious variety of formats ranging from those requiring a simple True/False response to the most complex where even the instructions seem to constitute a form of intelligence test, particularly on first encounter. As the fundamental purpose of an MCQ paper is to accurately and reliably assess factual knowledge in a particular subject, rather than measure innate intelligence or mental agility, it seems quite undesirable to knowingly introduce into the test any non-specific component which would adversely affect the validity of this method of examination.

MCQ format adopted in this book.

The question format adopted in the main is the "five choice

completion" which requires one response from a selection of five alternatives. There are of course many other types which are meritorious but as their individual components almost always can be reduced to a common basic True/False response for an all round appraisal as a combined test and teaching vehicle the simple one from five has no serious rival. A large number of the questions are posed in the negative form, presenting four correct items of information and only one wrong statement, thus more profitably imparting useful facts. A few examples of other MCQ types that may be met are given in Part Four.

The questions are set on alternate pages with the appropriate answer and commentary entered on the reverse of the page for convenience of consultation.

Self-assessment.

Provided that some adjustment may be required to correct disparity in curriculum content and course timing in different schools most medical students on completion of their course should score about 60 per cent while the best students should reasonably expect to score over 75 per cent.

Most marking schemes allot one mark for a correct answer and deduct one third or one half mark as a penalty for a wrong answer to discourage blind guessing. In an actual examination it is imperative to know if a countermark system is employed as it would be profitable to guess if no penalty were incurred when wrong.

Part One

MICROBIAL BIOLOGY AND HOST DEFENCES

1. Historical

1.1 Which one of these workers was *NOT* responsible for any of the following contributions to medical microbiology: discovery of lysozyme; discovery of specific bacteriolysis by immune sera; discovery of streptomycin; introduction of specific serodiagnosis of infection?

A. Fleming
B. Pasteur
C. Pfeiffer
D. Waksman
E. Widal.

1.2 Which one of these workers was *NOT* responsible for any of the following contributions to medical microbiology: first epidemiological study of cholera; rules for attributing aetiological role of bacteria in disease; introduction of prophylactic immunization; epidemiology and prevention of puerperal sepsis?

F. Jenner
G. Koch
H. Metchnikoff
J. Semmelweis
K. Snow.

1.3 Who proposed the "Clonal selection theory of acquired immunity"?

L. Burnet
M. Ehrlich
N. Jerne
O. Marrack
P. Medawar.

Answers overleaf

1.1 B. Although *Pasteur* has been called the "father of microbiology", he did not make any of the discoveries listed. *Fleming* discovered lysozyme in animal tissues and secretions (e.g. egg-white, tears) and later, of course penicillin. *Pfeiffer* demonstrated that cholera bacilli lysed when injected into the peritoneal cavity of an immune guinea pig, due to the combined action of antibody plus complement; *Waksman* discovered streptomycin which revolutionized the treatment of tuberculosis; *Widal* introduced his serological test, based upon tube agglutination of known strains of salmonellae by the patient's serum.

1.2 H. *Metchnikoff* was responsible for the concept of cellular (phagocytic) immunity. *Jenner*, observing that milk-maids working with cows infected with cowpox were resistant to smallpox, introduced artificial active immunization with cowpox lymph. *Koch* introduced solid media, pure culture plate techniques, elucidated the nature of tuberculosis and formulated "Koch's Postulates". *Semmelweis* recognized indirect spread of streptococcal puerperal fever from cadaver to patient on the hands of the attendant. *Snow* traced a local epidemic of cholera to the Broad Street pump in London and terminated the epidemic by removing the pump handle.

1.3 L. The theory claims that antibody-producing cells exist as a mixed population capable of forming antibody against the complete spectrum of antigenic configurations, but that each cell can synthesize antibody of only *one* specificity. Exposure of this mixed population to an antigen stimulates multiplication of cells which produce antibody closely corresponding to the particular antigen. This results in emergence of a family of genetically identical cells — *a clone* — which produces antibody of high specificity for the antigen which elicited the response. *Ehrlich* proposed the "side chain theory" of antibody formation and introduced acid-fast staining, arsenical chemotherapy for syphilis, and methods for toxin — antitoxin standardization; *Jerne* proposed a "natural selection theory" of antibody formation; *Marrack* proposed the "lattice theory" of antigen — antibody aggregation; *Medawar* proved the immunological basis of allograft rejection.

2. Bacterial Cytology

2.1 **Which organelle is ABSENT from the procaryotic cell?**

A. Flagellum
B. Mesosome
C. Nuclear membrane
D. Protoplasmic (plasma) membrane
E. Ribosome.

2.2 **Which material is present in both fungi and mycoplasmas but absent from bacteria?**

F. Chitin
G. Chlorophyll
H. Lipid
J. Muramic acid
K. Sterol.

2.3 **Which property will *NOT* differentiate fungi from bacteria?**

L. Formation of aerial hyphae in culture
M. Possession of mitochondria
N. Presence of nuclear membrane
O. Production of sexual spores
P. Synthesis of cell wall chitin.

2.4 **Which component of the bacterial cell contains alternating units of N-acetyl-D-glucosamine and N-acetylmuramic acid in a B-(1-4)-linkage?**

Q. Endotoxin
R. Flagella
S. Mesosome
T. Mucopeptide
U. Pili.

Answers overleaf

2.1 C. The procaryotic cell is much smaller and simpler in structure than the eucaryotic cell. Thus bacteria lack mitochondria, Golgi apparatus, endoplasmic reticulum and a nuclear membrane. Bacteria divide by binary fission and therefore do not show mitosis.

2.2 K. Sterols are characteristically present in cell membranes of eucaryotic cells (e.g. fungi) but absent from procaryotic cells (e.g. bacteria) with the exception of mycoplasmas, whose membrane they stabilize against osmotic lysis. Chitin is found in fungi; muramic acid is unique to bacteria; chlorophyll is found rarely in photosynthetic bacteria.

2.3 L. Cultures of nocardia and streptomyces sometimes appear powdery due to the formation of aerial hyphae, a feature shared with many fungi. The other properties are possessed by fungi and other eucaryotic cells but absent from bacteria.

2.4 T. The muramyl residues of the mucopeptide are frequently cross-linked by tetrapeptides of L-alanine, D-alanine, D-glutamic acid and L-lysine (or meso-diaminopimelic acid (DAP) and in some Gram-positive species by a further five glycine residues — the so-called pentaglycine bridge. The N-acetylmuramyl-tetrapeptide polymerises to form a large polyamine network which has been compared to a string vest. It provides a rigid mechanical support for the enclosed osmotically-fragile protoplast.

2.5 Which component of the bacterial cell wall is responsible for protection of the delicate protoplast from the effects of the high internal osmotic pressure?

V. Capsule
W. Endotoxin
X. Mucopeptide
Y. Somatic antigen
Z. Teichoic acids.

2.6 Which cell wall component does *NOT* form part of the lipopolysaccharide layer in Gram-negative bacilli?

A. Basal core polysaccharide
B. Endotoxin
C. Lipid A
D. Muramic acid
E. Somatic antigen determinants.

2.7 Which one is *NOT* true of bacterial capsules?

F. Act as virulence factors
G. May be demonstrated by either positive or negative staining
H. Most are composed of polysaccharide
J. Serial culture promotes capsulation
K. Swell with specific antisera.

Answers overleaf

2.5 X. The mucopeptide layer is a large polymeric network of alternating units of N-acetyl-D-glucosamine and N-acetyl muramic acid in a B-(1-4)-linkage cross-linked by tetrapeptide chains, that invests the osmotically-sensitive protoplast like a corset. In Gram-positives this layer forms the greater part of the cell wall, but in the Gram-negatives it forms only about 10 per cent, surrounded by a complex plastic layer of lipopolysaccharide.
The thick mucopeptide layer of Gram-positives has to withstand the internal osmotic pressure of 20 atmospheres whereas the thin R-layer of Gram-negatives has merely to withstand about 5 atmospheres. When the functional integrity of this layer is damaged, as by lysozyme or the action of penicillin and other cell wall antibiotics, the cells become deformed, swell and may rupture. Gram-negatives thus damaged tend to survive as spheroplasts, due to investment with the somatic antigen and the lesser osmotic effect.

2.6 D. Muramic acid is not a component of the lipopolysaccharide layer but part of the rigid basal cell wall, the R-layer of Gram-negatives.
N-acetyl muramic acid, together with N-acetyl glucosamine are the two repeating residues of the backbone polymer which when cross-linked by small peptides form the mucopeptide skeletal network. Outside this is a pliable layer of lipopolysaccharide which is probably not covalently bound to the rigid layer.

2.7 J. Repeated subculture selects non-capsulate mutants whereas animal passage selects fully capsulate, virulent forms. Some species have polypeptide, (anthrax bacillus) lipopolysaccharide (some Gram-negatives) or hyaluronic acid (streptococci) capsules. Apparent swelling of the capsule on contact with type-specific antisera forms the basis of the Neufeld or quellung reaction for serotyping.

2.8 Which organism forms an antiphagocytic capsule composed of a polypeptide of D-glutamic acid?

L. *Bacillus anthracis*
M. *Cryptococcus neoformans*
N. *Haemophilus influenzae type b*
O. *Klebsiella pneumoniae*
P. *Streptococcus pneumoniae type 3.*

2.9 Which organelle is ABSENT from all Gram-positive species of bacteria?

Q. Capsule
R. Endospore
S. Mesosome
T. Pili
U. Volutin granules.

2.10 Which has little or no ability to generate energy?

V. Bacteria in exponential phase of growth
W. Bacteria in lag phase of growth
X. Bacterial protoplast
Y. Bacterial spheroplast
Z. Ungerminated bacterial endospore.

2.11 With what specific cytological event do you associate the synthesis of dipicolinic acid?

A. Cell division
B. Cell separation
C. DNA replication
D. Protoplast formation
E. Sporulation.

2.12 Which step comes FOURTH in the preparation of a Gram film?

F. Acetone/alcohol
G. Carbol fuchsin
H. Crystal violet
J. Fix by flaming
K. Gram's iodine.

Answers overleaf

2.8 L. All the listed organisms develop prominent capsules but only *B. anthracis* forms one composed of polymeric D-glutamic acid, the others all forming polysaccharide capsules.
In all cases, the capsular material acts as an aggressin or virulance factor, by neutralization of opsonins and inhibition of phagocytosis.

2.9 T. Common holdfast pili and the larger F pilus which acts as a conjugation tube between mating bacterial cells have not been demonstrated in Gram-positive species which are incapable of genetic exchange by the mechanism of conjugation.

2.10 Z. Prior to germination the spore is dormant and biochemically inert with no evidence of respiratory or biosynthetic activity whereas the other answers relate to bacterial forms capable of energy transformation. The exponential or logarithmic growth phase shows the most active stage of multiplication but even the lag phase has considerable biochemical activity.

2.11 E. Spores contain a relatively high concentration of a calcium-chelating chemical, dipicolinic acid, which is absent from all vegetative bacteria. It is thought the presence of Ca-dipicolinate in the spore cortex contributes to heat resistance.
Germination is accompanied by release of Ca-dipicolinate into the medium, with concomitant loss of heat resistance and a loss of refractility.

2.12 F. A Gram-stained bacterial smear is prepared in the following sequence: 1. Fix by flaming; 2. Crystal violet; 3. Gram's iodine; 4. Decolourise by acetone/alcohol mixture;
5. Counter stain with carbol fuchsin. Gram-positive is blue, Gram-negative is pink.

3. Bacterial Physiology and Genetics

3.1 Which one of the following would *NOT* be found in a pure bacterial culture?

A. Capsulated and non-capsulated cells
B. Cells of more than one strain
C. Cells with different flagellar antigens
D. Smooth and rough colonies
E. Spores and vegetative cells.

3.2 Which technique allows enumeration of living bacteria in a liquid culture?

F. Assay of cell mass by nephelometry
G. Estimation of DNA concentration
H. Microscopic count
J. Miles-Misra plate count
K. Weighing of total cell mass.

3.3 Which one is characteristic of the exponential (log) growth phase?

L. High resistance to cell-wall antibiotics
M. Low oxygen consumption
N. Synchronous division
O. Unbalanced growth
P. Viable count is close to total count.

3.4 Which is most closely associated with the stationary phase of growth?

Q. Depletion of critical metabolite
R. Exponential growth
S. Greatest oxygen consumption
T. Metabolic inactivity
U. Shortest generation time.

Answers overleaf

3.1 B. By definition, a pure culture must contain only one strain.
The other answers are false, as in every case the variant forms can arise as a result of mutation, selection and phenotypic variation. Sporulation is dependent upon the stage of growth and maturation of the individual cell and it is usual to find mixtures of vegetative and sporing organisms in a pure culture of clostridium or anthracoid.
To obtain a pure culture it is necessary to pick out a typical colony and sow it on a new medium. By repeating this process once or twice it is usually possible to separate the colony forming unit from possible contaminating organisms of a different strain.

3.2 J. This technique measures only the number of colony-forming units and is called a "viable count". Error arises from multiplication without separation (e.g. streptococci) aggregation after separation (e.g. mycobacteria and corynebacteria) and from technical errors.

3.3 P. During the exponential phase the number of dead or dormant cells is negligible and the entire culture is actively, although not synchronously, dividing. This requires a high oxygen consumption for active metabolic function and growth is balanced with the growing cell walls at their most vulnerable to cell wall active antibiotics (e.g. penicillin).

3.4 Q. Bacterial populations cease to increase either when they exhaust one of the ingredients essential for growth or when some metabolite or a combination of metabolites, reaches an inhibitory concentration in the medium. It is most unlikely that simple overcrowding is of any significance in limiting growth in liquid culture.

3.5 Absence of which enzyme from most anaerobic bacteria accounts for the lethal effect of oxygen on strict anaerobes?

V. Catalase
W. Coagulase
X. Oxidase
Y. Hyaluronidase
Z. Urease.

3.6 Which factor inhibits the germination of tetanus spores in the tissues?

A. Calcium ions introduced with contaminating soil
B. Devitalized and necrotic tissue
C. High local oxidation-reduction potential
D. Presence of foreign bodies
E. Toxins of contaminating *Clostridium welchii.*

3.7 For what reason is penicillin lethal to bacteria yet essentially non-toxic for mammalian cells?

F. High population density of host cells neutralize its effect
G. Mammalian cells are impermeable to penicillin
H. Mammalian cells inactivate penicillin
J. Penicillin does not affect cells with long generation time
K. Target site is unique to bacteria.

3.8 Which statement about the action of penicillin is FALSE?

L. Gram-negative bacilli are usually resistant
M. Interferes with normal murein synthesis
N. Mammalian cells unaffected
O. Most effective against resting organisms
P. Serious hypersensitivity reactions may occur.

3.9 Which one of the following constituents is ABSENT from Lowenstein-Jensen medium used for the cultivation of mycobacteria?

Q. Agar
R. Eggs
S. Glycerol
T. Malachite green
U. Potato starch.

Answers overleaf

3.5 V. Although anaerobic organisms cannot utilize oxygen as a terminal electron acceptor they do possess the early part of the respiratory chain of enzymes through which hydrogen is combined with oxygen to form H_2O_2. Most aerobic organisms contain the enzyme catalase which degrades H_2O_2 into H_2 and O_2, but anaerobes generally lack catalase which allows the accumulation of H_2O_2 to a lethal concentration.

3.6 C. High oxidation-reduction potential would prolong dormancy. All the other factors listed decrease the oxidation-reduction potential and promote germination.

3.7 K. Penicillin acts by interfering with the formation of the rigid cell wall mucopeptide which is unique to bacteria. Mammalian cells lack this target component which gives penicillin a high degree of selective toxicity for bacteria. Harmful effects of penicillin for man are due to its ability to act as a hapten, thus inducing hypersensitivity.

3.8 O. Penicillin inhibits cross-linkage of glycopeptide during synthesis but is without effect on existing murein. The susceptible rigid layer of Gram-negative species is often protected by outer layers of lipopolysaccharide. Mammalian cells uniformly lack murein substrate. Penicillin and its degradation products may bind to proteins and act as allergens.

3.9 Q. L—J slopes are prepared by inspissating egg — mineral salts — potato starch — asparagine — Malachite green and glycerol. The coagulated egg protein provides a solid medium without any need for agar.

3.10 For what reason are fluid enrichment media used in clinical microbiology?

V. To dilute inhibitory substances in the specimen
W. To prevent multiplication during transit
X. To provide essential nutrients for fastidious species
Y. To selectively cultivate pathogens originally few in number
Z. To stimulate the growth of all enteric bacteria.

3.11 Which organism is most likely to be resistant to penicillin G?

A. *Clostridium tetani*
B. *Neisseria meningitidis*
C. *Staphylococcus aureus*
D. *Streptococcus pyogenes*
E. *Treponema pallidum.*

Answers overleaf

3.10 Y. Enrichment media are used when it is required to isolate a specific pathogen from large numbers of other bacteria present in the same specimen which might otherwise prevent detection of the pathogen. Such conditions are met in the standard examination of faeces, for intestinal pathogens of the salmonella or shigella genera. By inoculating with faeces a broth medium containing selenite (or brilliant green) commensal *Escherichia coli* is suppressed and any salmonellae or shigellae present multiply without inhibition thus producing an absolute increase in their numbers compared with the inhibited *E. coli*. On subculture to a solid medium the enteric pathogen will form a greater proportion of the inoculum and will therefore develop sufficient colonies to ensure its detection and successful isolation. When such a selective agent is added to a solid medium it again aids the isolation of the pathogen by partial inhibition of contaminating organisms but it does not increase the numbers of the pathogen as estimated by numbers of colony forming units.

By combining these two methods it is possible first to obtain an absolute increase in the pathogen by culture in an enrichment broth and then to improve its isolation and selection by subculture to a solid selective medium from which isolation in pure culture can be made.

3.11 C. With the exception of *Staphylococcus aureus* all the other organisms are likely to be penicillin sensitive.

In the early days of penicillin most strains of *Staph. aureus* were also sensitive but due to widespread use of the drug sensitive strains have been greatly reduced and replaced by resistant strains by a process called emergence of resistance. It is now known that the ability to produce penicillinase can be transferred by transduction, often involving transmission of a plasmid.

3.12 Which one of the following is the active agent in bacterial transformation?

F. Chemical mutagen
G. Episome
H. Extracted or released DNA
J. Temperate bacteriophage
K. Virulent bacteriophage.

3.13 Which one of the following when added to a culture of non-encapsulated pneumococci derived from a type 1 strain will result in biosynthesis of type 3 polysaccharide?

L. Bacteriophage specific for type 3 strain
M. DNA extracted from virulent type 3 strain
N. Killed rough mutants from type 3 strain
O. Specific opsonin against type 3 strain
P. Type 3 polysaccharide.

3.14 In bacterial transformation, how much of the donor genome is usually integrated with the DNA of the recipient cell?

Q. About half
R. All of it
S. A small fraction
T. Most of it
U. None of it.

Answers overleaf

3.12 H. Griffiths first described transformation during his studies on *Strep. pneumoniae*. He injected 3 groups of mice; the first with living non-capsulated cells; the second with dead capsulated cells; the third with living non-capsulated cells plus killed capsulated cells. The first two groups were unaffected but the third group died. He later recovered capsulated cells from the dead mice.
Avery, McLeod and McCarty (1943) identified the transforming agent as DNA. Transformation results from transfer of DNA containing the specific gene, absent from the living non-capsulated organism, which controls capsule formation.

3.13 M. This is genetic transfer by naked DNA transformation. Fragments of donor DNA containing about 20 genes pass through the cell wall and become integrated with the DNA of competent recipient cells which then acquire the genetic trait coded for in the fragment, in this example, biosynthesis of polysaccharide type 3.

3.14 S. When the donor cell disintegrates, its circular genome breaks into approximately 100 pieces, each containing about 20 genes. Recipient cells which are *competent* accept one of these fragments which is only about 1 per cent of the donor genome. Different cells will be transformed in different ways depending upon which pieces of donor DNA they receive so that the frequency of transformation for any one gene is usually about 1 per cent of the recipient population.

3.15 By what mechanism does β-phage control toxin production in *Corynebacterium diphtheriae?*

V. Conjugation
W. Lysogenic conversion
X. Random mutation
Y. Transduction
Z. Transformation.

3.16 Which one of the following is an example of lysogenic conversion?

A. Elimination of temperate phage
B. Induction of the lytic phase
C. Production of immunity to superinfection
D. Prophage controlled toxogenicity
E. Replication of virulent phage.

3.17 Which description applies specifically to Hfr cells?

F. Cells capable of conjugation
G. Cells with F pili
H. Cells with integrated F factor
J. F^+ cells
K. Donor cells in genetic transfer.

Answers overleaf

3.15 W. *C. diphtheriae* can synthesize its protein exotoxin only if infected by a temperate bacteriophage (β-phage). Whereas virulent phage replicate and cause lysis, temperate phage can either do the same or integrate with the DNA of the host bacterium and replicate with the bacterium as a prophage, without lysis.
It is in this phase that the temperate phage exerts its control over some of the functions of the cell, a phenomenon termed lysogenic conversion. Any strain not lysogenized is non-toxigenic and therefore avirulent.

3.16 D. A bacterium which carries a non-infectious form of phage is said to be lysogenic, and the phage termed prophage or temperate phage. Rarely they do exert a controlling influence on specific biochemical activities of the host cell, and when this happens the process is termed lysogenic conversion.
Important examples of lysogenic conversion are the control of diphtheria and scarlatina toxins and alterations in somatic antigen specificities.

3.17 H. Any bacterial cell which carries an F (fertility) factor is termed F^+ and those which lack it, F^-. The F^+ cells are capable of transferring their F factor to F^- recipient cells by conjugation via the sex pilus produced by the action of the F factor. This facilitates the transfer of the F factor to F^- cells which become F^+. The F factor exists as an extrachromosomal piece of DNA or plasmid in most cells but in a few, only one in 10^5 cells, it is integrated with the host DNA. Such cells have the ability to transfer the entire chromosome which results in recombination. They are termed Hfr strains, for high frequency of recombination, and their progeny are also Hfr cells.

3.18 What statement does <u>*NOT*</u> apply to bacterial plasmids?

- **L.** Associated with genetic transfer by conjugation
- **M.** Exist as self-replicating extra-chromosomal DNA
- **N.** Found only in Gram-negative species
- **O.** May be transferred between different genera
- **P.** Sometimes confer multiple antibiotic resistance.

Answers overleaf

3.18 N. This statement is false as plasmids have been described in many Gram-positives, although in these, transfer can only be achieved by transduction as conjugation is restricted to Gram-negative species.
By definition they never integrate with the nucleus of the cell although a plasmid in one cell may become an integrated episome in another.

4. Host-Parasite Relationships

4.1 Which term best denotes the relatively harmless bacteria which inhabit the surfaces of the skin, upper respiratory tract and intestine?

A. Commensals
B. Invasive organisms
C. Non-invasive organisms
D. Obligate pathogens
E. Saprophytes.

4.2 Which site normally lacks a commensal flora?

F. Appendix
G. External auditory meatus
H. Female lower genital tract
J. Nasal vestibule
K. Trachea.

4.3 Which one of the following genera predominates in the initial colonization of the lower intestine in healthy breast-fed babies?

L. *Escherichia*
M. *Clostridium*
N. *Lactobacillus*
O. *Staphylococcus*
P. *Streptococcus.*

Answers overleaf

4.1 A. Commensals are relatively harmless parasitic bacteria which by selection and adaptation have succeeded in colonizing the epithelial surfaces of the skin, gut, upper respiratory tract and female lower genital tract. By their presence they prevent an overgrowth of more harmful organisms which are thus reduced to the role of opportunistic pathogens, and they help limit the establishment of obligate pathogens to those with enhanced powers of infectivity. There is a characteristic though not constant flora in the different regions which reflects local differences in the microbial environment. Many of the commensals are potential pathogens in that they can assume a pathogenic role if the surface epithelial defences are diminished by disease or injury or if they are transported to a site other than their normal habitat. Examples of commensal species are *E. coli, Strep. viridans, Staph. albus, Cl. welchii.*

4.2 K. The respiratory tract, in health, is virtually sterile below the larynx. Any organism entering the lower respiratory tract tends to be trapped on the mucus coating of the ciliated epithelium. It is then wafted upwards in secretions to be swallowed or expectorated.
The other sites normally possess their own specific commensal flora.

4.3 N. At birth the intestinal tract is sterile but rapidly becomes colonized by bacteria ingested in food. After weaning, and earlier in bottle fed babies, Gram-negative bacilli, enterococci and clostridia predominate in the commensal flora.

4.4 Which bacterial group forms over 90 per cent of the commensal flora of the intestine in an adult?

Q. Facultative anaerobic coliforms
R. Gram-positive bacilli
S. Obligate aerobes
T. Obligate anaerobes
U. Obligate pathogens.

4.5 Which organism has a specific affinity for ciliated epithelium?

V. *Bordetella pertussis*
W. *Haemophilus parainfluenzae*
X. *Listeria monocytogenes*
Y. *Mycoplasma pneumoniae*
Z. *Streptococcus pneumoniae.*

4.6 Which term BEST describes the clinical syndrome which is associated with a rapid, continuous invasion of the blood stream by bacteria?

A. Bacterial endocarditis
B. Invasive infection
C. Septicaemia
D. Toxaemia
E. Transient bacteraemia.

Answers overleaf

4.4 T. Normal human faeces contains approximately 10^{11} bacteria per gram, of which over 90 per cent are obligate anaerobic Gram-negative bacilli of the bacteroides — fusiformis group, with *B. fragilis* the predominant species.

4.5 V. *Bordetella pertussis* shows a highly specific organotropism for the bronchial ciliated epithelium which might explain its localization in whooping cough. This specific predilection for ciliated epithelium has been studied experimentally in the chick embryo. Infection of 15 day embryos localized to the ciliated epithelium of the respiratory tract whereas similar challenge of a 12 day old embryo, in which ciliated epithelium has not yet developed, did not cause infection of the respiratory tract.

4.6 C. Septicaemia is a clinical syndrome associated with severe bacteraemic infection characterized by frequent or continuous invasion of the blood stream from one or more foci of infection in the tissues. There is usually a degree of toxaemia which in serious Gram-negative infections may progress to endotoxic or septicaemic shock.
Bacteria which enter the blood stream do not continue to circulate long enough to multiply, but are rapidly cleared by splenic and hepatic macrophages. This has been demonstrated experimentally in a case of septicaemia in which surgical drainage of the inciting local septic focus, an abscess, resulted in concomitant clearing of the blood stream.

4.7 What term is used for any infection which is accompanied by the signs and symptoms of disease?

F. Clinical infection
G. Febrile illness
H. Invasive infection
J. Latent infection
K. Nosocomial infection.

4.8 Which one of these commensal organisms is MOST likely to become active as an opportunistic pathogen during prolonged treatment with a broad-spectrum antibiotic?

L. *Actinomyces israeli*
M. *Candida albicans*
N. *Haemophilus influenzae*
O. *Streptococcus pneumoniae*
P. *Streptococcus viridans.*

Answers overleaf

4.7 F. An infection which produces symptoms as evidence of disease is said to be a clinical, overt or apparent infection as distinct from a subclinical, latent or inapparent infection where clinical signs of disease are absent.
Sub-clinical infection is important in the development of natural active immunity. Repeated or continued latent infection stimulates the immune response without the production of disease, or it may persist as the carrier state after resolution of a clinical infection.
The term nosocomial infection is applied to hospital acquired infection and this includes hospital cross infection.

4.8 M. The protective effect of microbial competition exerted by the commensal flora prevents unrestricted growth of *Candida albicans* in the mouth, intestine and female lower genital tract. Prolonged treatment with any broad spectrum antibiotic markedly depletes the normal flora with subsequent emergence or outgrowth of resistant species. *Candida albicans*, being unaffected by antibacterial antibiotics may rapidly become dominant in colonization of the mucous membranes and is then prone to invade the tissues with the production of thrush or progressive infection.

5. Mechanisms of Pathogenicity

5.1 **Which microbial property does *NOT* interfere with the body's attempt to localize infection?**

A. Fibrinolysin or *Streptococcus pyogenes*
B. Motility of *Salmonella paratyphi B*
C. Polysaccharide capsule of *Streptococcus pneumoniae*
D. Protein A of *Staphylococcus aureus*
E. Vi antigen of *Salmonella typhi.*

5.2 **Which one of the properties listed here as a characteristic of bacterial exotoxins is FALSE?**

F. Diffusible, highly toxic, immunogenic protein
G. Marked specificity of action
H. May mimic disease when injected
J. Neutralized by specific IgG antibody
K. Produced only by Gram-positive species.

5.3 **Which one of the following statements about the nature of diphtheria toxin is FALSE?**

L. Inhibits protein synthesis in target cells
M. Is a slow acting, heat labile, diffusible protein exotoxin
N. Production greatly influenced by iron concentration
O. Toxigenic cells are invariably lysogenic
P. Toxoid is derived from avirulent strains.

Answers overleaf

5.1 B. Motility seems to have no influence on invasiveness as is shown by absence of any protection afforded by homologous anti-flagellar antibody and by the finding that non-flagellate variants of invasive pathogens show no related loss of virulence or invasiveness.

All the other properties listed contribute to virulence and invasiveness. Bacterial capsules, Vi antigen and protein A all inhibit opsonization and phagocytosis thereby allowing organisms to multiply and spread instead of being killed at the site of infection.

Fibrinogen, which escapes from blood vessels in and around the focus of infection, is normally rapidly changed to fibrin to enmesh and localize organisms, but this effect is prevented by fibrinolysin which acts as an aggressin by dissolving fibrin.

5.2 K. Although formed predominantly by Gram-positive species as typified by *Cl. tetanus* (tetanus), *C. diphtheria* (diphtheria) and *Strep. pyogenes* (scarlet fever), important exceptions among the Gram-negative species include *Sh. dysenteriae* (dysentery), *V. cholerae* (Asiatic cholera) and a limited range of "O" serotypes of *E. coli* (not infantile gastroenteritis but a benign cholera-like disease).

5.3 P. Diphtheria toxoid is prepared from a potent toxin produced by a *virulent* strain of *C. diphtheriae*. The toxin is treated with dilute formalin until all toxicity is lost, while antigenicity and serological specificity are retained. Toxoid can be injected to stimulate the formation of specific antitoxin in active immunization.

5.4 Which one of the following shows if a clinical isolate of *Corynebacterium diphtheriae* is toxigenic?

Q. Bacterial agglutination with specific antisera
R. Colonial morphology on blood tellurite medium
S. Positive Elek plate
T. Positive identification of "gravis" strain
U. Positive Schick test.

5.5 Which organism produces a potent exotoxin?

V. *Shigella boydi*
W. *Shigella dysenteriae*
X. *Shigella flexneri*
Y. *Shigella sonnei*
Z. None of these.

5.6 Which description of tetanus toxin is FALSE?

A. Blocks synaptic inhibitory transmitter
B. Converted to toxoid by formalin
C. Formed only by virulent strains of *Clostridium tetani*
D. Reaches CNS via motor nerves
E. Six serologically distinct forms are known.

5.7 Which organism produces a toxin which blocks neuromuscular transmission in cholinergic motor nerves?

F. *Clostridium botulinum*
G. *Clostridium tetani*
H. *Corynebacterium diphtheriae*
J. *Shigella dysenteriae*
K. *Vibrio cholerae.*

Answers overleaf

5.4 S. Identification of a clinical isolate as *C. diphtheriae* of a certain serotype or biotype says nothing of its possible toxin producing capacity; a "gravis" strain could be non-toxigenic and therefore non-virulent. The Schick test is quite inappropriate as it is not a test of toxigenicity but a skin test to detect individuals lacking protective immunity. The Elek plate is therefore what is used in most laboratories, although some still use animal toxicity tests and others use a more recently introduced test based on toxicity for tissue culture HeLa cells. The Elek plate method is a double diffusion precipitation-in-gel where the test strain and control strains, are streaked at right angles to a paper strip impregnated with diphtheria antitoxin. A positive result shows precipitation lines of identity with the test and positive control strains.

5.5 W. The severe diarrhoea and profound dehydration which characterize *S. dysenteriae* infection is due largely to the effect of a heat stable, protein exotoxin lethal to intestinal epithelial cells. Early literature described it as a "neuro-toxin". The other species do not produce an exotoxin but possess the common ability to penetrate epithelial cells and to release endotoxin on lysis.

5.6 E. Although different strains of *Clostridium tetani* show diversity in somatic and flagellar antigens, all toxigenic strains produce only one type of tetanospasmin with a common serological specificity.

5.7 F. The neurotoxin of *Clostridium botulinum* suppresses the release of acetylcholine at the myoneural junction of cholinergic motor nerves first producing muscle weakness and then paralysis, usually fatal if it involves the respiratory muscles.

5.8 Which organism exerts its pathogenic effect by secretion of a potent exotoxin which continuously stimulates the adenyl cyclase activity of mucosal cells?

L. *Clostridium welchii*
M. *Salmonella typhimurium*
N. *Shigella dysenteriae*
O. *Staphylococcus aureus*
P. *Vibrio cholerae*

5.9 Which one of the following is a property of endotoxin from coliform bacilli?

Q. Completely neutralized by antisera
R. Firmly cell-bound and non-diffusible
S. Largely composed of protein
T. Purified endotoxin is strongly antigenic
U. Toxicity lost by boiling for 1 hour.

5.10 "Cord formation" is characteristically shown by cultures of virulent strains of which organism?

V. *Bordetella pertussis*
W. *Clostridium tetani*
X. *Corynebacterium diphtheriae*
Y. *Haemophilus influenzae*
Z. *Mycobacterium tuberculosis.*

5.11 Which property of *Mycobacterium tuberculosis* is thought to determine its destructive effect in the body?

A. Ability to induce delayed hypersensitivity
B. Ability to produce "cord factor"
C. Classical endotoxic activity
D. Classical exotoxic activity
E. Potential for intracellular growth.

Answers overleaf

5.8 P. Water and electrolyte transport across mucosal cells is normally controlled by intermittent hormonal activation of the mucosal enzyme adenyl cyclase which then catalyses the formation of cyclic adenosine monophosphate (CAMP) from ATP and this activates the cell. Cholera exotoxin causes continuous irreversible activation of adenyl cyclase with resultant "rice water" stools and lethal water and electrolyte depletion.

5.9 R. Endotoxin is a constituent of the Gram-negative cell wall lipopolysaccharide or somatic antigen. Endotoxin activity is most closely associated with the lipid A component. It is normally part of the cell wall and non-diffusible, but may be released as macromolecular disintegration products on lysis of the cell.
These complexes are only weakly antigenic and do not induce fully protective antibodies. They are relatively heat stable. They contrast with bacterial exotoxins which are pharmacologically active diffusible protein poisons of high specificity, highly antigenic, fully neutralized by specific anti-serum, heat labile and produced mainly by Gram-positive organisms.

5.10 Z. Virulent strains of *Mycobacterium tuberculosis* regularly form in culture serpentine cords of longitudinally arranged, pallisaded bundles of strongly acid-fast bacilli, whereas avirulent strains lack this characteristic. "Cord formation" is related to the production of a toxic glycolipid, trehalose-6, 6′-dimycolate, which is known as "cord factor".

5.11 A. Although the full mechanism of pathogenicity is still unknown it seems probable that immunologically active surface peptidoglycolipid induces a state of delayed hypersensitivity to the bacillary proteins. Once induced, small amounts of bacillary antigen will enhance resistance but large amounts have a distinctly destructive effect due to an exaggerated level of tuberculin hypersensitivity. "Cord factor" may contribute to local tissue damage but it is not an important factor.

5.12 In which of the following is bacteraemia LEAST likely to occur?

F. Anthrax
G. Brucellosis
H. Diphtheria
J. Miliary tuberculosis
K. Subacute bacterial endocarditis.

5.13 Metastatic pyaemic abscesses occur MOST frequently as a complication of infection with which organism?

L. *Brucella abortus*
M. *Corynebacterium diphtheriae*
N. *Salmonella typhimurium*
O. *Staphylococcus aureus*
P. *Streptococcus viridans.*

Answers overleaf

5.12 H. Loeffler, who established the aetiological agent of diphtheria in 1884, made the important discovery that even in fatal cases the organism remained localized to the tissue damage at the portal of entry and could never be isolated from the distant lesions in the heart, adrenals, liver and nervous tissue. He postulated the existence of a diffusible toxin which is now known to be the essential mechanism of virulence of *C. diphtheriae*. Diphtheria and tetanus are often cited as the two best examples of exotoxic disease.

Anthrax and brucellosis are characteristically septicaemic. Miliary tuberculosis is due to systemic dissemination of tubercle bacilli through the circulation and may be likened to septicaemic pyaemic abscesses. Bacterial endocarditis is characterized by multiple arterial emboli often containing clumps of bacteria in the fragments of clot. All of these infections, with the exception of tuberculosis, may therefore be diagnosed by blood culture, and even in the case of tuberculosis, Lowenstein claimed to make positive isolation of *M. tuberculosis* by blood culture although the work could not be repeated by Jensen.

5.13 O. *Staphylococcus aureus* infections are characterized by suppuration and a tendency to bacteraemic spread. Even minor superficial skin lesions can result in the development of visceral and skeletal abscesses, due to haematogenous spread. Thus minor boils may give rise to major systemic infections, such as acute osteitis, acute bacterial endocarditis, lung, brain and perinephric abscesses, all of which may produce a fatal septicaemia.

Of the other organisms listed, *C. diphtheriae* is virtually non-invasive; *S. typhimurium* not infrequently enters the blood stream but does not produce systemic abscesses as would the enteric fever and septicaemic salmonellae. *Strep. viridans* may cause subacute bacterial endocarditis with an embolic bacteraemia but does not produce pyaemic abscesses; *Brucella abortus* generally invades the circulation, but granulomata rather than abscesses are found at its site of lodgement.

6. Host-Defence Mechanisms

6.1 Which one of the following is an example of innate or natural immunity?

A. Natural recovery by crisis in pneumococcal pneumonia
B. Neonatal immunity to haemophilus meningitis
C. Prophylactic immunization with diphtheria toxoid
D. Protection afforded by tetanus antitoxin
E. Resistance in early stages of a primary infection.

6.2 Which property is *NOT* included in the non-specific immune response?

F. Chemotactic factors
G. Interferon
H. Lysozyme
J. Memory cells
K. Phagocytosis.

Answers overleaf

6.1 E. Innate or natural immunity is the resistance shown on the first encounter between a particular host species and a particular species of parasite, before the forces of acquired immunity are mobilized. The basis of this immunity is not clear but includes the mechanism of non-specific resistance as well as "natural antibodies". Their absence from germ-free animals suggest they are a form of acquired immunity to cross reacting antigens.

The other answers are all examples of acquired immunity. Natural recovery in lobar pneumonia when defervescence dramatically takes place by crisis from 7th—10th day is due to the sudden acquisition of protective levels of pneumococcal antibody by natural active immunity. Neonatal resistance to *H. influenzae* is dependent upon transplacental transfer of protective maternal antibody, an example of natural passive immunity. Tetanus antitoxin produces immediate but short-lived artificial passive immunity whereas diphtheria toxoid produces artificial active immunity which is slow to develop but of long duration.

6.2 J. Memory and specificity are characteristics of the specific immune response which emphasize its striking difference from the non-specific response. In the latter the response to an initial encounter with a foreign antigen (and indeed irritants of non-antigenic nature) is repeated in an identical way on each subsequent encounter so there is no progressive augmentation of the reaction to afford an enhanced resistance to the challenge.

The specific response differs in showing a different reaction on the first or primary response from the subsequent or secondary response. This is due to the initial proliferation and differentiation of specifically reactive cells which persist until a subsequent encounter, when they respond more rapidly and in greater number. Likewise the cell-mediated response is augmented by rapid proliferation of specifically reactive cells surviving from the primary response.

6.3 Which is *NOT* a property of interferon?

L. Acts extracellularly to inactivate virus
M. Coded for by cellular DNA
N. Not virus-specific but is cell specific
O. Released from infected cells
P. Synthesis triggered by double-stranded RNA.

6.4 Which one of the following does *NOT* contribute to non-specific immunity?

Q. Ciliated respiratory epithelium
R. Commensal bacterial flora
S. IgE
T. Neutrophil chemotaxis
U. Skin fatty acids.

6.5 Which cell is an essential constituent of pus?

V. Basophil
W. Eosinophil
X. Erythrocyte
Y. Lymphocyte
Z. Neutrophil.

Answers overleaf

6.3 L. Interferons act intracellularly to inhibit the replication of viruses in contrast to antibodies which act extracellularly to inactive virus. Interferons are cell specific in production and activity, but differ in size, antigenic specificity and activity according to the animal species of the producer cell, even when induced by the same virus.

6.4 S. Elaboration of the IgE class of antibody results from specific antigenic stimulation of certain plasma cells found mainly in the linings of the gut and respiratory tract and is therefore a component of the specific immune response. Reaginic antibody, as it is called, can initiate allergic reactions in its three target organs, the gut, skin and respiratory tract. Union of an allergen with cell bound IgE causes release of vasoactive substances which are active in the allergic response. Clinical conditions which are caused by IgE include food allergy with vomiting, diarrhoea and abdominal pain, atopic eczema, hay fever and bronchial asthma.
The IgE molecule has a sedimentation value of 8S, molecular weight of 196,000 and a very short half-life of 1.5 days, and is normally present in only trace amounts in the serum.

6.5 Z. Neutrophils make up 70 per cent of the white cells of peripheral blood where they function as the principal phagocytic cell of acute inflammation. They escape from inflamed capillaries at a focus of infection, attracted by chemotactic factors, to engulf and digest micro-organisms in the tissues. If the challenge is strong, in numbers or virulence, many polymorphs are injured or killed with release of autolytic enzymes from their lysosomes. Thus in a pyogenic infection, collections of dead and dying polymorphs are always present in an established lesion as a creamy semi-fluid pus.

6.6 **Which one of the following does *NOT* relate to the non-specific defence enzyme, lysozyme?**

A. Absent from normal urine, CSF and sweat
B. Classified as a muramidase
C. Found in tears, saliva and nasal secretions
D. Mainly active against Gram-negative bacteria
E. Protoplasts, spheroplasts and L-forms are not susceptible to its action.

6.7 **Which is the MOST abundant immunoglobulin class in a normal healthy adult?**

F. IgA
G. IgD
H. IgE
J. IgG
K. IgM

6.8 **Which antibody naturally crosses the placental barrier in man?**

L. IgA
M. IgD
N. IgE
O. IgG
P. IgM

Answers overleaf

6.6 D. The lytic effect of lysozyme is dependent upon free access of the sensitive linkages in the mucopeptide layer of the cell wall in both Gram-positive and Gram-negative bacteria. In Gram-positives this layer is often exposed and therefore vulnerable to attack so they are quite commonly sensitive, whereas in Gram-negatives, the equivalent R-layer is submerged in the total cell wall protected by the overlying thick lipopolysaccharide complex of somatic antigen and endotoxin. Combined treatment of Gram-negatives with complement plus antibody against the somatic antigen opens to attack the hidden but essentially sensitive mucopeptide, so that lysozyme can then contribute to the total bactericidal effect.
Cell wall deficient forms of bacteria lack the sensitive R-layer and are completely resistant to the action unless they revert to the normal bacterial form.

6.7 J. IgG class contains the bulk of protective antibody against infecting agents which elaborate potent protein exotoxins or act as extra-cellular parasites or at least exhibit an extracellular phase during dissemination to target cells. This protection results from neutralization of toxins, promotion of phagocytosis by opsonic action, bactericidal effect of antibody plus complement and viral neutralization. The normal mean serum concentrations of the different immunoglobulin classes in mg per 100 ml are as follows: IgA 280; IgD 3; IgE 0.03; IgG 1240; IgM 120.

6.8 O. Prior to birth the fetus forms little or no antibody primarily because of the lack of antigenic stimulation due to the protective function of the placenta, but also to some extent because of immunological immaturity. Passive protection is afforded by placental transfer of maternal IgG against a wide variety of infections. This is an example of natural passive immunity.

6.9 Which is the MOST prominent immunoglobin class present in external secretions?

Q. IgA
R. IgD
S. IgE
T. IgG
U. IgM.

6.10 Which property attributed to IgM is FALSE?

V. Combines with specific antibody
W. Fixes complement
X. Functions as protective antibody
Y. Most abundant antibody class in uninfected newborn
Z. Specific attachment to phagocytes.

6.11 Which class of immunoglobulin is MOST effective in specific bacterial agglutination?

A. IgA
B. IgD
C. IgE
D. IgG
E. IgM.

Answers overleaf

6.9 Q. IgA comprises 10 per cent of serum immunoglobin, second only to IgG. It is produced mainly in response to local stimulation of subepithelial lymphoid tissues of the respiratory, intestinal and genito-urinary tracts, serving an important antimicrobial function in their external secretions by neutralizing viruses and inhibiting epithelial adhesions of bacteria. IgA does not fix complement or cross the placenta.
In serum, it occurs as a 7S monomer with two heavy and two light chains. In secretions, it is combined with a protein secretory or transport piece, which makes it resistant to proteolytic enzymes, and occurs mainly as dimers (9S) trimers (11S) and even bigger combinations.

6.10 Y. Serum antibody in the newborn is almost exclusively of maternal origin as only IgG is able to cross the placenta. The presence of significant levels of IgM in the newborn is evidence of fetal synthesis in response to infection *in utero*, (e.g. high rubella IgM antibody in fetal blood in congenital rubella).

6.11 E. Specific bacterial agglutination results from the combination of antibodies with specific surface antigens on the bacterial cell wall to form bridges between adjacent bacterial cells, causing them to clump together. The relative ability of the different immunoglobulin species to achieve this result is dependent upon their capacity to form bridges rather than to unite with single cells.
The IgM molecule is unique in possessing five radiating arms with separate antigen binding sites and therefore is easily the most effective agglutinin.
IgG is more efficient in forming a large lattice with soluble antigen and is therefore a better precipitin than IgM.

6.12 Which is an example of artificial active immunity?

F. Administration of pertussis vaccine
G. Neonatal resistance to diphtheria
H. Prophylactic immunization with anti-tetanus serum
J. Subclinical attack of brucellosis
K. Treatment of diphtheria with antitoxin.

6.13 How does the secondary immune response differ from the primary response with respect to antibody production?

L. Antibody produced is of low specificity
M. Less antibody is produced more quickly
N. Less antibody is produced more slowly
O. More antibody is produced more quickly
P. More antibody is produced more slowly.

6.14 Which term refers to a substance which when mixed with antigen will enhance the antibody response by eliciting a higher and more prolonged production of antibody to the antigen?

Q. Adjuvant
R. Chemotactic factor
S. Complement
T. Hapten
U. Immunogen.

Answers overleaf

6.12 F. Artificial active immunity is produced by the administration of an antigenic preparation in order to raise humoral or cell-mediated immunity so as to increase acquired resistance to the natural disease. For exotoxic diseases such as tetanus or diphtheria the antigen is the relevant toxoid, and for invasive or intracellular infections, vaccines of living attenuated organisms (e.g. BCG against tuberculosis) or killed virulent organisms (e.g. TAB against enteric fever or pertussis against whooping cough) are employed.

6.13 O. The secondary response is characterized by the production of elevated levels of specific antibody, predominantly of 7S IgG class by the phenomenon of "recall production". The second encounter with a particular antigen stimulates the proliferation and differentiation of plasma cells from memory cells formed during the primary response. This augmented response affords an increasing level of protective immunity against many bacterial pathogens and may be acquired artificially by repeated "booster" doses of antigen or naturally by recurring clinical or subclinical infection.

6.14 Q. Adjuvants may be incorporated into a vaccine to improve the level and duration of response. They include not only a wide range of chemical and biological agents but also the physical presentation of the antigen (e.g. water-in-oil emulsion). Some adjuvants are suitable for use in humans (e.g. alum-precipitated diphtheria toxoid) whereas others are unsuitable due to local or general toxicity which stimulate a marked granulomatous reaction (e.g. water-in-oil emulsion with or without the addition of killed tubercle bacilli as in Freund's complete and incomplete adjuvant.)

6.15 Why are children highly susceptible to a wide range of so-called "childhood fevers" which are rare in the adult?

V. A child's skin is very delicate and not protective
W. Acute inflammation is an adult response
X. Childhood infections cause immunological tolerance
Y. Children cannot produce protective antibody
Z. Natural active immunity is only gradually acquired.

6.16 Which one of the following antibodies produced in response to natural infection has NO protective value in either arresting infection or preventing clinical disease?

A. *Corynebacterium diphtheriae* anti-toxin
B. *Haemophilus influenzae* capsular antibody
C. *Streptococcus pneumoniae* type specific opsonin
D. *Streptococcus pyogenes* M-protein antibody
E. *Treponema pallidum* Wassermann-type antibody.

6.17 Which one of the following specific immune mechanisms offers NO protection against infection?

F. Immunological tolerance to the pathogen
G. Lytic effect of antibody plus complement
H. Neutralization of protein exotoxin
J. Opsonic action of antibody
K. Viral neutralization by specific antibody.

Answers overleaf

6.15 Z. The natural acquisition of active immunity to an increasing spectrum of organisms results from repeated contact with specific and related antigens throughout life. The child is born with virtually no protective active immunity and as a result of sub-clinical and clinical infection gradually builds an effective level of immunity to commonly occurring organisms. The incidence of such childhood infections is particularly high in day nurseries, kindergartens and in the primary school due to the prevalence of initial encounters with pathogens.
The human can produce IgM at birth, IgA by 3 weeks and IgG after 4—6 weeks. It is current practice to immunize the newborn with BCG in areas where tuberculosis is prevalent.

6.16 E. Wassermann-type antibody is a 19S class "reagin" (don't confuse with IgE reaginic antibody) or autoantibody directed against tissue-derived antigens. The level of WR antibody is an expression of the extent of tissue damage but does not influence or correlate with the degree of resistance to reinfection.

6.17 F. The development of tolerance to microbial antigens which reach the fetus *in-utero* before maturation of the immune system can completely inhibit future protective specific immune response to the same or closely related parasites. Thus instead of enhancing resistance, tolerance will increase susceptibility to progressive infection or will allow establishment of prolonged latent infection with organisms of low virulence.
The other answers all refer to components of the specific immune response which are active in protection against infection.

6.18 The term "immunological enhancement" has a specific meaning. Which one of the following refers to this phenomenon?

L. Accelerated graft rejection by second set reaction.
M. Enhanced resistance effected by booster immunization
N. Hyperimmunization by repeated antigenic stimulation
O. Secondary or anamnestic antibody response
P. Specific antibody-induced suppression of graft rejection.

6.19 Which lymphoid tissue is rarely if ever the site of germinal centres and antibody production in the absence of disordered function?

Q. Lymph node
R. Peyer's patches
S. Spleen
T. Thymus
U. Tonsil.

6.20 When serum from a sensitized donor is injected intradermally into a non-sensitized recipient and after a latent period greater than one day an injection of the same antigen into the same site elicits within minutes localized weal and flare reaction, which class of antibody is specifically associated with the reaction?

V. IgA
W. IgD
X. IgE
Y. IgG
Z. IgM.

Answers overleaf

6.18 P. Ignorance of the specific usage of the term immunological enhancement would result in rejection of the one correct answer, in favour of four spurious correct answers. Immunological enhancement refers to the prolongation of the survival time of a tumour allograft as a result of prior immunization with killed tumour cells or antitumour serum. The effect is attributed to soluble complexes of antibody and tumour antigen masking sites either on the tumour or on the surface of the host lymphocytes, preventing the induction or expression of cell-mediated immunity to the tumour.

6.19 T. Germinal centres, plasma cells, gammaglobulins and specific antibody are found in the thymus only in association with autoimmune diseases or lymphoma, whereas they are typically present in the other tissues.
The thymus provides the environment in which T-lymphocytes mature and proliferate in preparation for their role in killing foreign or aberrant cells. During their working life these T-cells reside not in the thymus but in the thymus dependent areas of peripheral lymphoid tissues, or they recirculate in the blood and lymph.
Congenital thymus aplasia (Di George syndrome) is a rare non-hereditary developmental abnormality which results in selective deficiency of T-cells and cell-mediated reactions. Combined thymus dysplasia and agammaglobulinaemia (Swiss type) is an X-linked or autosomal recessive inherited deficiency of both cell-mediated and humoral immunity.

6.20 X. This is the Prausnitz-Küstner reaction which detects by passive transfer cytophilic or cytotropic IgE responsible for atopic hypersensitivity.
The P-K reaction is now very rarely used clinically due to the danger of transmitting serum hepatitis. Total and specific serum IgE can now be estimated *in vitro* by radioimmunoassay, using labelled anti-IgE.

6.21 Which pharmacologically active mediator of hypersensitivity is *NOT* involved in Type I immediate hypersensitivity reactions of classical anaphylaxis?

A. Bradykinin
B. Histamine
C. Lymph node permeability factor (LNPF)
D. Serotonin or 5-hydroxytryptamine (5 HT)
E. "Slow-reacting substance" of anaphylaxis (SRS-A)

6.22 Which statement about the pathogenesis of acute immune-complex disease (serum sickness) is FALSE?

F. Antigen-antibody complexes circulate in the blood stream
G. Immune aggregates are deposited in blood vessels and kidney glomeruli
H. Immune complexes fix complement causing release of neutrophil chemotactic factors
J. Tissue damage is mediated by complement and neutrophils
K. Such Type III hypersensitivity reactions occur only on second or subsequent encounters with an antigen.

Answers overleaf

6.21 C. Lymph node permeability factor is released by sensitized T-lymphocytes upon their interaction with the sensitizing antigen in a Type IV delayed hypersensitivity reaction (e.g. tuberculin reaction). It is only one of a number of lymphokines released by reactive lymphocytes in the delayed response (e.g. macrophage inhibition factor (MIF), mitogenic factor, lymphotoxin, chemotactic factors etc.). The other vasoactive agents are the principal mediators of Type I immediate hypersensitivity reaction tissue damage. They are released by mast cells and basophils of sensitized individuals when injected antigen reacts specifically with cytophilic IgE antibody bound to the cell surface. This reaction is the basis of classical anaphylaxis and is independent of complement which cannot be fixed by IgE antibody. Release of vasoactive substances is associated with degranulation of the mast cells as the active agents are contained in the mast cell granules.

6.22 K. Serum sickness typically follows the first large injection of foreign protein (e.g. diphtheria antitoxin produced in the horse) when new antibody combines with antigen remaining in the circulation to produce small soluble immune complexes, formed in antigen excess, which are retained in the blood stream. Interaction of antigen with cytophilic IgE on basophils releases vasoactive amines, causing increased vascular permeability (Type I reaction). This allows escape of the smaller complexes while trapping the larger ones in the basement membrane of vessels and in kidney glomeruli. These immune aggregates activate the complement sequence causing tissue damage and release of chemotactic factors. Neutrophil polymorphs attracted by chemotaxis to vascular lesions (rarely to glomeruli) adhere to immune complexes (immune adherence by C3) and attempt to eliminate them by phagocytosis. This causes release by neutrophil hydrolytic enzymes which are mediators of the local Arthus-type response. Symptoms which appear after 10 days — 2 weeks from primary injection include fever, rashes, lymph node and splenic enlargement, pain and swelling of joints and glomerulonephritis.

6.23 Which one of the following is *NOT* a manifestation of delayed hypersensitivity?

L. A positive Mantoux reaction
M. Contact hypersensitivity to penicillin
N. Inhibition of macrophage migration by specific antigen
O. Local Arthus reaction
P. Skin graft rejection.

6.24 Which term refers to a graft from a genetically dissimilar donor of the same species?

Q. Allograft
R. Autograft
S. Isograft
T. Syngraft (more commonly, syngeneic graft)
U. Xenograft.

Answers overleaf

6.23 O. The Arthus reaction is an antibody mediated hypersensitivity reaction produced e.g. at the site of injection of antigen into a sensitized recipient. The antibodies responsible are IgG or IgM which precipitate with the antigen to form complexes in or around blood vessels.
The Ag-Ab complex fixes complement and when phagocytosed by polymorphs causes the release of cathepsin with the production of necrosis. Clinically the reaction occurs within 1 hour of the injection, becoming maximal by 6—12 hours before subsiding.

6.24 Q. Survival of tissues transplanted into an immunologically competent individual is greatly influenced by the closeness of genetic identity which determines histocompatability. Closest identity is found in an *autograft* from one site to another on the same individual. An almost identical situation pertains with an *isograft* or *syngeneic graft* which are synonomous terms for grafts between identical twins or inbred animals. The Isologous tissue has almost 100 per cent compatability with the recipient's tissues, provided they are of the same sex.
Grafts between genetically dissimilar members of the same species are termed *allografts* and the tissue is said to be homologous. This is the usual type of clinical graft to replace damaged or diseased human tissues (skin grafts and other organ transplants between non-identical humans). The fetus may be considered a natural allograft.
A graft between members of different species (e.g. man and apes) is termed a *xenograft* and the tissue said to be heterologous. Differences in tissue antigens are great and rejection is rapid and complete.

6.25 Which one is *NOT* a test of phagocytic function?

V. Basophil degranulation test
W. Measurement of specific leucocyte enzymes
X. Quantitative nitroblue tetrazolium dye reduction
Y. Phagocytic index
Z. Test of bactericidal activity.

6.26 Which one of the following is *NOT* a hypersensitivity reaction?

A. Allergen skin tests in assessment of hay fever
B. Contact dermatitis
C. Schick test
D. Tuberculin reaction
E. Urticaria following a penicillin reaction.

Answers overleaf

6.25 V. The basophil (mast cell) degranulation test is not related to neutrophil phagocytic function as are the other tests but is used to detect IgE antibody. Antigen injected into a sensitized animal combines with IgE attached to basophils to cause the release of vasoactive substances, including histamine and serotonin, which are mediators of acute inflammation.

The other tests are all used to detect deficiencies in the capacity of neutrophils to ingest and kill bacteria.

The phenomenon of phagocytosis involves a sequential chain of events including directional chemotaxis, the actual ingestion process by either surface or opsonic phagocytosis, formation of a phagosome and its subsequent union with a granular lysosome to form a phagolysosome within which the acid hydrolases kill and digest the engulfed organism. By measuring the individual components of the process it is possible to identify the nature of a specific defect in this vital non-specific defence mechanism.

6.26 C. The Schick test is based upon *in vivo* neutralization of an intradermal injection of diphtheria toxin by circulating antitoxin of IgG class and is unrelated to any form of hypersensitivity mechanism.

Allergen skin tests in hay fever detect specific cell-bound IgE to such allergens as pollen.

Contact dermatitis is due to delayed hypersensitivity acquired through contact with allergens which combine with tissue proteins to form haptens. These may be detected by patch tests.

Urticaria of immunological type is a type of immediate hypersensitivity mediated by the release of histamine and other vasoactive substances due to combination of the allergen with IgE or due to the formation of antigen-antibody complexes.

7. Characteristics of Viruses

7.1 Which one is <u>*NOT*</u> true of virus reproduction?

A. They derive their energy from the host cell
B. They reproduce only within living cells
C. They remain fully organized and infectious throughout replication
D. They use host-cell synthetic mechanisms to form virus protein
E. Virus-specific protein may inhibit host-cell DNA synthesis.

7.2 Which one of the following is NEVER suitable for cultivation of viruses?

F. Chorio-allantoic membrane of the chick embryo
G. Continuous cell lines such as HeLa cells
H. Human erythrocytes from susceptible individuals
J. Suckling mice
K. Viable bacterial culture.

7.3 Which virus produces a "steady-state" infection with release of virus particles by surface "budding" and delayed host-cell lysis?

L. Echovirus
M. Influenza virus
N. Poliomyelitis virus
O. Polyoma virus
P. Simian virus 40 (SV40).

Answers overleaf

7.1 C. Viruses do not reproduce as complete independent units as do rickettsiae and bacteria. Instead, the virus nucleic acid replicates at a different site in the cell from that where virus proteins are synthesized and it is only at the very beginning and end of the replication cycle that complete virions exist within the infected cell.

7.2 H. Viruses are obligate intracellular parasites which can multiply only in susceptible living cells. Human erythrocytes lack the necessary nucleus and normal cytoplasmic organelles and cannot support the growth of viruses.

7.3 M. Virus-cell relationships are essentially of three types, examples of each being included in the answers.
The "*steady-state*" relationship is characterized by development and maturation of virus particles near the cell membrane and their slow release by "budding". The cell usually survives for some time before lysis finally occurs (e.g. influenza virus). The *cytolytic* cell response results from rapid intracellular multiplication and assembly of virus particles which are released on lysis of the host cell (e.g. poliomyelitis and echoviruses). The *integrated* interaction occurs with tumour viruses which multiply synchronously with the host cell which may also undergo a productive cytolytic infection (e.g. SV40 and polyoma virus).

7.4 Which one of the following groups of viruses characteristically causes the production of a reverse transcriptase to allow persistence of the viral genome in all transformed cells?

Q. Oncogenic adenoviruses
R. Oncogenic herpesviruses
S. Oncogenic poxviruses
T. Oncornaviruses
U. Papovaviruses.

7.5 Which one of the following viruses possesses a lipid-rich envelope and is therefore ether sensitive?

V. Adenoviruses
W. Herpesviruses
X. Papovaviruses
Y. Picornaviruses
Z. Reoviruses.

7.6 Which group of viruses are usually isolated in primate tissue cultures incubated at 33°C?

A. Cytomegaolvirus
B. Mumps virus
C. Poliovirus
D. Rhinovirus
E. Smallpox virus.

Answers overleaf

7.4 T. Viral oncogenesis is characterized by persistence of the viral genome in each transformed cell. With DNA-containing tumour viruses (e.g. papovaviruses, oncogenic adenoviruses and herpesviruses) the virus needs only to become integrated with the host cell DNA but this is not directly possible with the oncornaviruses which first require an extra step. By possessing *reverse transcriptase* (RNA-dependent DNA polymerase) the virus first has its RNA genome transcribed into the DNA form which is then integrated with the host DNA. Thus it can subsequently code for its own synthesis in each transformed daughter cell. It is of interest that similar RNA-dependent DNA polymerase enzymes have been identified in virus particles recovered from human milk in families with a high incidence of breast cancer.
The other viruses are all DNA-containing tumour virus groups which neither have nor need a reverse transcriptase.

7.5 W. Ether resistance is a characteristic of all the listed viruses with the exception of herpesvirus. The high lipid content of certain viruses is related to the presence of a lipid-rich envelope derived from the host-cell membrane.

7.6 D. Rhinoviruses grow much more readily if incubated at the reduced temperature of 33°C which is similar to the temperature of the human nasopharynx. The other viruses listed all prefer to be incubated at 36°—37°C.
Cytomegalovirus is limited in cell culture to human fibroblasts. Mumps virus can be grown in cell culture, producing characteristic multinucleated giant cells or in the amniotic sac of the chick embryo. Poliovirus can be cultivated in primary or continuous cell lines of human or monkey origins, with production of a cytopathic effect. Smallpox virus can be isolated either in tissue culture or in the chorioallantoic membrane of the chick embryo, with the development of characteristic pocks.

7.7 Which one of the following is present in rickettsiae but *NOT* in viruses?

F. DNA and RNA (both present)
G. Double stranded DNA
H. Double stranded RNA
J. Single stranded DNA
K. Single stranded RNA.

7.8 What is *NOT* present in bacteria?

L. Both RNA and DNA
M. Double stranded DNA
N. Double stranded RNA
O. Single stranded DNA
P. Single stranded RNA.

7.9 What is the most sensitive and specific method for detection and measurement of hepatitis B antigen (HB_SAG or Australia antigen)?

Q. Complement fixation test
R. Counter electrophoresis
S. Immunodiffusion in gel
T. Radioimmunoassay
U. Standard tube precipitin reaction.

Answers overleaf

7.7 F. Each virus particle (virion) contains a single type of nucleic acid, either DNA or RNA, but never both. This is an important property which distinguishes viruses from any form of cell; cells always contain both.

7.8 N. Double-stranded RNA has so far been demonstrated only in reoviruses (respiratory enteric orphan viruses) and certain plant viruses.

7.9 T. Radioimmunoassay is an extremely sensitive technique for accurately measuring antigens in very low concentrations. Several methods are available but that most commonly used is a commercial RIA sandwich type test where antibody coated plastic beads are incubated with patient's serum. After washing, the beads are then incubated with antibody labelled ^{125}I. After further washing the bound radioactivity is measured in a gamma counter. The test can be done in a few hours.

8. Disinfection and Sterilization

8.1 **Which term is applied to chemicals used to kill bacteria on an inanimate object?**

A. Antibiotic
B. Antiseptic
C. Chemotherapeutic agent
D. Disinfectant
E. Sterilizing agent.

8.2 **Which chemical disinfectant is incorporated into many antiseptic soaps used for pre-operative hand-washing?**

F. Cetrimide
G. Chlorhexidine
H. Crystal violet dye
J. Hexachlorophane
K. Mercuric chloride.

8.3 **Which statement about the chemical disinfectant Cetrimide (e.g. Cetavlon) is FALSE?**

L. Chemically is a quaternary ammonium compound
M. Does not kill spores
N. Greatly inactivated by organic matter
O. Totally inactivated by soap
P. Wide antibacterial spectrum with bactericidal activity.

Answers overleaf

8.1 D. *Disinfection* is any chemical treatment of an inanimate object with the intention of killing organisms. Disinfectants are influenced by concentration, time, nature of the micro-organisms present, presence of protein which might protect the organisms or presence of organic matter which might inactivate the chemical. Spores and bacteria with a high lipid content (e.g. tubercle bacilli) are generally most resistant.

Antisepsis is restricted to methods suitable for treatment of living tissues such as skin, open wounds, the membrane linings of the body cavities and certain hollow organs.

The use of these terms depends entirely upon convention: some authors do not distinguish between them and simply apply the general term disinfection to both situations.

Sterilization is any process which removes (i.e. filtration) or kills (e.g. heat, irradiation, etc) all forms of life. It is therefore "all or none".

8.2 J. Hexachlorophane is a chlorinated bi-phenol. It has a detergent action which disrupts the cell membrane and exerts a cumulative bacteriostatic activity. It is used in soaps, bath sachets and concentrates to suppress the growth of skin bacteria. It is also incorporated into deodorants to suppress commensal organisms which decompose sweat.

8.3 P. Cetrimide has a narrow range of activity and is more bacteriostatic than bactericidal. As a cationic detergent it is useful for disinfection of surfaces and can be used for skin antisepsis as in Cetavlon or in combination with Hibitane in Savlon.

8.4 Which type of chemical disinfectant is effective against Gram-positive bacteria but is ineffective against Gram-negative species?

Q.	Chloroxylenol	(e.g. Dettol)
R.	Hypochlorites	(e.g. Chloros)
S.	Iodine	
T.	Iodophores	(e.g. Betadine)
U.	Phenolics	(e.g. Clearsol).

8.5 Which of the following physical agencies is LEAST effective in killing bacteria of most genera suitably exposed to its effect?

V. Desiccation by lyophilization
W. Dry heat (160°C for 1 hour)
X. Ionizing radiations (β-rays), γ-rays and X-rays)
Y. Moist heat (121°C for 15 minutes)
Z. Shortwave UV radiation (2,500 Å).

8.6 What pressure, temperature and time combination is satisfactory for standard, autoclave usage?

A.	0 lb/in^2	100°C	20 min
B.	0 lb/in^2	100°C	60 min
C.	15 lb/in^2	121°C	20 min
D.	15 lb/in^2	121°C	60 min
E.	low vacuum	80°C	20 min.

8.7 What are the correct conditions for sterilization of glassware in a hot air oven?

F. 60°C for 1 hour
G. 100°C for 30 minutes on 3 successive days
H. 121°C for 20 minutes at a pressure of 15 lbs/in^2
J. 160°C for 1 to 2 hours
K. 500°C for 20 minutes.

Answers overleaf

8.4 Q. With the exception of Chloroxylenol all these disinfectants have a wide antibacterial spectrum. Chloroxylenols and hypochlorites are greatly inactivated by organic matter. Particularly resistant species such as *Ps. aeruginosa* and other pseudomonads have been found contaminating Chloroxylenol solutions.

8.5 V. This is the process of freeze-drying used to *preserve* stock cultures of bacteria in a viable but dormant state. Organisms drying from the fluid phase, such as exhalation droplet-nuclei formation, show a high initial death rate which is greatly reduced by drying from the frozen state as in lyophilization.
The other physical agents are all lethal for bacteria given suitable exposure. Heat is used for laboratory sterilization, beta and gamma irradiation for sterilization of plastic and rubber articles and X-rays mainly for research on their biological effect. UV irradiation is used in safety cabinets for handling hazardous materials such as tubercle cultures and occasionally for reducing aerial transmission of infection in wards and operating theatres.

8.6 C. 121°C (15 lb/in^2) for 20 minutes ensures the death of all spores, the most heat resistant of life forms. Increasing the exposure time to 60 minutes would simply damage relatively heat sensitive materials such as rubber gloves and culture media.

8.7 J. Sterilization by dry heat requires a higher temperature maintained for a longer time than the temperature-time ratio for steam under pressure. The use of the very high temperature of 500°C is unnecessary and might damage glassware.

8.8 Which statement about ethylene oxide sterilization is FALSE?

L. Efficiency tested by exposure of spore strip in chamber
M. Kills bacteria by alkylating nitrogenous components
N. Non-harmful to man and chemically inert in air
O. Process requires controlled humidity, temperature and pressure
P. Used to sterilize heat sensitive instruments.

8.9 Which is a satisfactory method for the sterilization of antisera before use?

Q. Boil for 5 minutes
R. Inactivate at 56°C for 30 minutes
S. Pass through 0.2 μ membrane filter
T. Steam at atmospheric pressure for 1 hour
U. Steam at 15 lb pressure for 20 minutes.

Answers overleaf

8.8 N. Even small amounts of residual chemical remaining on a sterilized endoscope before use has a powerful toxic vesicant action on the tissues. The gas is highly explosive and to limit this reactivity it is usually provided in small canisters combined with 90 per cent CO_2 or a fluorocarbon. *Bacillus subtilis var globigi* spore strips are the recommended biological test method.

8.9 S. Filtration through a bacterial filter will sterilize antisera without loss of potency. Steam at 15 lb pressure for 20 minutes is standard autoclave procedure for heat resistant articles but would result in loss of activity of antibody and is therefore unsuitable. Heating at 56°C for 30 minutes would merely inactivate complement.

Part Two

MICRO-ORGANISMS OF MEDICAL IMPORTANCE

1. Pyogenic Cocci

1.1 Which is a specific character of *Staphylococcus aureus?*

A. Coagulase
B. Golden pigment
C. Haemolysin
D. Phosphatase
E. Salt resistance.

1.2 Which biological property is possessed by *Staphylococcus aureus* but lacking from *Staphylococcus albus?*

F. Ability to grow on MacConkey plate
G. Catalase activity
H. Mannitol fermentation
J. Penicillinase production
K. Protein A production.

1.3 The ability to produce which one of the following distinguishes *Staphylococcus aureus* from *Staphylococcus albus?*

L. Beta-lactamase
M. Coagulase
N. Gelatinase
O. Pigment
P. Phosphatase.

Answers overleaf

1.1 A. Coagulase is a diffusible protein produced by pathogenic staphylococci, which by definition may then be called *Staph. aureus* or *Staph. pyogenes*. Demonstration of its production is taken as the most reliable criterion of human pathogenicity.
In the presence of a plasma co-factor, prothrombin, coagulase converts soluble fibrinogen into insoluble fibrin, which in the tube test for its activity appears as a jelly-like clot. Its role in virulence is debatable. By its action in the tissues it deposits a protective barrier around the organisms which might interfere with phagocytosis thus protecting the parasite, or conversely, it may surround the focus with a fibrin network which effectively localizes the infection by encapsulation of the lesion.

1.2 K. Of the listed properties Protein A production alone is restricted to *Staph. aureus* where it is present as a surface component on most strains but is absent from a few strains of proven virulence. Protein A has the unique ability to bind to the Fc region of IgG and may be shown to have a role in pathogenesis, as it interferes with opsonization and is actively antiphagocytic.

1.3 M. Free coagulase or the cellbound "clumping factor" are generally accepted as being exclusive properties of *Staph. aureus*, uniformly absent from all *Staph. albus* strains. Coagulase is also held to be the best correlate of human pathogenicity.

1.4 Which is MOST likely to kill any random strain of *Staphylococcus aureus?*

Q. Ampicillin
R. Crystal violet dye
S. Desiccation
T. Lysozyme
U. 10 per cent sodium chloride.

1.5 To which lytic phage group do *Staphylococcus aureus* strains responsible for toxic-epidermal-necrolysis belong?

V. Group I
W. Group II
X. Group III
Y. Group IV
Z. Other.

1.6 Which antibiotic is active against hospital strains of *Staphylococcus aureus?*

A. Ampicillin
B. Amoxycillin
C. Carbenicillin
D. Methicillin
E. Penicillin G.

1.7 For which genus has the phosphatase test been used to help identify human pathogens?

F. *Corynebacterium*
G. *Haemophilus*
H. *Leptospira*
J. *Neisseria*
K. *Staphylococcus.*

1.8 Which disease does *NOT* result from infection by a Gram-positive coccus?

L. Impetigo
M. Rheumatic fever
N. Scalded skin syndrome
O. Staphylococcal food poisoning
P. *Streptococcus viridans* subacute bacterial endocarditis.

Answers overleaf

1.4 R. Although *Staph. aureus* is relatively more resistant to physical and chemical agents than most other vegetative organisms it is highly sensitive to crystal violet which is strongly bactericidal. Crystal violet may be incorporated in a blood agar plate when it is expected that a heavy growth of staphylococci might mask the presence of a small but significant number of *Strep. pyogenes* which is resistant to this dye.

1.5 W. Phage group II (e.g. phage type 71) produce an exfoliative toxin which causes skin to peel off in sheets, giving the appearance of scalded skin. (Staphylococcal cross infection in hospital is most often caused by strains of groups I and III — e.g. phage types 29, 52, 52A, 79, 80 and 42E, 47, 53, 54, 83A.) Type 77 strains resistant to methicillin are now being encountered.

1.6 D. Hospital strains of *Staphylococcus aureus* almost always produce penicillinase and Methicillin is the only antibiotic listed which is resistant to this inducible beta-lactamase.

1.7 K. There is a close correlation between phosphatase activity, coagulase production and virulence in *Staphylococcus aureus*. Other useful, though less reliable criteria include mannitol fermentation and gelatin liquefaction which are generally negative with non-pathogenic strains.

1.8 O. Staphylococcal food poisoning results from ingestion of *Staphylococcus aureus* enterotoxin produced in contaminated food prior to ingestion. It is therefore an intoxication but *not* an infection.
Impetigo usually results from mixed *Staphylococcus aureus* and *Streptococcus pyogenes* infection of the skin.
Rheumatic fever is a delayed allergic complication of any *Streptococcus pyogenes* infection. The "scalded-skin" syndrome (toxic-epidermal-necrolysis) is caused by *Staphylococcus aureus*, usually of phage group II. It mainly affects babies.

1.9 Which is the major virulence determinant of *Streptococcus pyogenes?*

Q. Group-specific carbohydrate
R. Hyaluronic acid capsule
S. Streptokinase
T. Streptolysin "O"
U. Type-specific M-protein.

1.10 Which antigen of *Streptococcus pyogenes* is both the major determinant of virulence and the principal stimulus for protective immunity against infection?

V. Erythrogenic toxin
W. Group A carbohydrate
X. Hyaluronidase
Y. M-protein
Z. Streptolysin "O".

1.11 Which antigen largely determines the adherence of *Streptococcus pyogenes* to oropharynx epithelium?

A. Erythrogenic toxin
B. Group-specific C-substance
C. Protoplast membrane
D. T-protein
E. Type-specific M-protein.

Answers overleaf

1.9 U. M-protein is the sole protective antigen and a prerequisite for virulence. It initiates colonization of the nasopharynx, an essential step in the establishment of infection, by virtue of its special adhesive affinity for the epithelial cell surface. The importance of this role is shown by the protective action of type-specific IgA which blocks adhesion and prevents infection. Furthermore, removal of M-protein by trypsin causes loss of adhesion, inability to initiate infection and therefore loss of virulence.

M-protein is strongly antiphagocytic, a property neutralized by circulating specific antibody which opsonizes the streptococcal surface. It also precipitates fibrin, which protects the organism by interfering with opsonization. Although it has no direct leucocidin effect it may damage polymorphs by the products of its toxic effect on platelets. The other properties listed all contribute to virulence by either direct toxic action, promotion of invasion or by an association with rheumatic lesions.

1.10 Y. M-protein is strongly antiphagocytic, toxic to platelets, determines adhesion to oropharyngeal mucosal cells and encourages establishment of infection. It stimulates the production of type specific protective antibody which can negate its various harmful effects. Strains lacking M-protein are either of reduced virulence or avirulent.

1.11 E. Production of specific secretory IgA antibody against a particular Griffiths type prevents lodgement by coating the M-protein thus preventing adherence which is the first stage in establishment of infection.

1.12 **Which biological product of *Streptococcus pyogenes* does *NOT* contribute to invasiveness?**

F. Erythrogenic toxin
G. Hyaluronic acid capsule
H. Hyaluronidase
J. M-protein
K. Streptokinase.

1.13 **Which statement is TRUE of only a few *Streptococcus pyogenes* serotypes?**

L. Causally related to acute glomerulonephritis
M. Causally related to acute rheumatic fever
N. Member of Lancefield group A
O. Sensitive to bacitracin
P. Sensitive to penicillin.

1.14 **Which is the inciting agent of acute rheumatic fever?**

Q. Any strain of beta-haemolytic streptococcus
R. Any streptococcus
S. Any streptococcus with C-substance antigen
T. Only Lancefield group A streptococci
U. Only specific serotypes of *Streptococcus pyogenes*.

1.15 **Which property is possessed by only a small number of serological types of *Streptococcus pyogenes?***

V. Ability to cause acute glomerulonephritis
W. Ability to cause acute rheumatic fever
X. Bacitracin sensitivity
Y. Lancefield group A determinant
Z. Susceptibility to penicillin.

Answers overleaf

1.12 F. With the exception of producing the rash of scarlet fever, erythrogenic toxin has no known role in streptococcal pathogenesis. All the other products contribute to invasiveness. The hyaluronic acid capsule is antiphagocytic; hyaluronidase hydrolyses the intercellular ground substance of tissues and has been termed a "spreading factor". Streptokinase activates the fibrinolytic system which digests fibrins and inhibits localization of the infection.

1.13 L. The relationship of streptococcal sore throat or skin infection to the allergic sequelae of acute glomerulonephritis is limited to a few common serotypes and occasional strains from some less common types. Type 12 (throat strain) and type 49 (throat and skin strain) are the commonest nephritogenic types. The pathogenesis is probably due to deposition of streptococcal membrane antigen immune complexes in the glomeruli.
The other statements are true of all types of *Streptococcus pyogenes.*

1.14 T. Rheumatic fever can follow infection by any strain of *Streptococcus pyogenes* (synonym for Lancefield group A streptococci) and it is suggested that cross reacting antigens exist in the streptococcal cell membrane and the cardiac sarcolemma as well as between the group specific C-substance and structural glycoprotein of the heart valves.

1.15 V. Nephritogenic strains of *Streptococcus pyogenes* are limited to those which possess a polypeptide component with antigen determinants in common with the glomerular membrane. Griffiths type 12 is the most prevalent but types 4, 25 and 49 also occur. This is essentially an immune-complex disease. Other diseases caused by *Streptococcus pyogenes*, including rheumatic fever, are not type specific although scarlet fever is associated with a restricted range of types.

1.16 Which property of *Streptococcus pyogenes* causes the rash of scarlet fever?

A. Deoxyribonuclease
B. Erythrogenic toxin
C. Hyaluronidase
D. Streptokinase
E. Streptolysin O.

1.17 Which will *NOT* distinguish *Streptococcus pneumoniae* from *Streptococcus pyogenes?*

F. Bile solubility test
G. Blood agar plate culture
H. Capsule swelling (quellung or Neufeld) reaction
J. Gram reaction and catalase test
K. Optochin sensitivity test.

1.18 Which test is NEVER of differential value in distinguishing *Streptococcus pneumoniae* from *Streptococcus viridans?*

L. Bile solubility
M. Catalase reaction
N. Colonial morphology
O. Optochin sensitivity
P. Quellung reaction.

1.19 Which organism may be easily demonstrable in direct Gram-films but will not grow on culture unless plates are inoculated promptly or the swab is transmitted to the laboratory in a transport medium?

Q. *Bacillus anthracis*
R. *Corynebacterium diphtheriae*
S. *Listeria monocytogenes*
T. *Mycobacterium leprae*
U. *Neisseria gonorrhoeae.*

Answers overleaf

1.16 B. The production of erythrogenic toxin is controlled by specific temperate bacteriophages (lysogenic conversion). Dick test negative reactors, with circulating antitoxin, show only the local septic effects of infection whereas Dick test positive reactors who lack antitoxic immunity, develop clinical scarlet fever. Immunity against erythrogenic toxin does not protect against streptococcal infection, immunity to which is determined by type-specific anti-M-protein antibody.

1.17 J. Both organisms are Gram-positive and catalase negative so these tests have no differential value. *Strep. pneumoniae* is bile soluble, optochin sensitive, reacts specifically with capsular antiserum and forms alpha haemolytic "draughtsman" colonies whereas *Strep. pyogenes* lacks these properties.

1.18 M. The catalase reaction is negative for all streptococci and has no differential value. The pneumococcus is bile soluble, forms characteristic "draughtsman" colonies, is sensitive to optochin (ethylhydrocupreine hydrochloride) and gives a positive quellung, or Neufeld capsule swelling reaction, with specific antiserum.

1.19 U. The viability of *Neisseria gonorrhoeae* diminishes rapidly with delay in setting up cultures, due to cooling and desiccation, unless protected by use of Stuart's transport medium, or Transgrow medium.
Myocobacterium leprae can neither be stained by Gram's method nor be cultivated in artificial media. It can be grown in immunologically deficient mice and hamsters.

1.20 Which statement about *Neisseria gonorrhoeae* is FALSE?

V. Avirulent strains lack endotoxic activity
W. Colonies are oxidase-positive irrespective of virulence
X. Fresh clinical isolates are usually piliated and virulent
Y. Repeated subculture results in loss of virulence
Z. Virulent strains have a surface K-like polysaccharide envleope.

1.21 Which observation about *Neisseria meningitidis* strongly supports the view that it is essentially an extracellular pathogen?

A. Anticapsular antibodies give group specific protection
B. Epidemic strains exist
C. Invasive qualities highly developed
D. Meningococcaemia commonly occurs
E. Typically intracellular within polymorphs.

1.22 Which one of the following is a genus composed of anaerobic Gram-negative diplococci?

F. *Acinetobacter*
G. *Moraxella*
H. *Neisseria*
J. *Peptococcus*
K. *Veillonella.*

Answers overleaf

1.20 V. Virulence seems to depend upon the adhesive property of pili and the antiphagocytic effect of K-antigen envelope. Endotoxins of P-nucleoprotein and somatic lipopolysaccharide types possessed equally by both virulent and avirulent strains probably have a contributory but not primary role in virulence.

1.21 A. The level of circulating specific antibody to extracellular pathogens (e.g. *Strep. pneumoniae; H. influenzae; Strep pyogenes*) correlates with the degree of resistance whereas intracellular pathogens (e.g. *Br. abortus; L. monocytogenes; M. tuberculosis; S. typhi*) show no similar correlation between humoral immunity and resistance.

1.22 K. *Veillonella* are obligate anaerobic Gram-negative diplococci. *Peptococcus* is the only other anaerobic group listed but is Gram-positive. All of these organisms, with the exception of *Peptococcus*, may have identical morphology in a Gram-film, appearing as Gram-negative diplococci.

2. Gram-positive Bacilli

2.1 Which one of the following is NOT a property of *Corynebacterium diphtheriae?*

A. "Chinese letters" arrangement of cells
B. Gram-positive bacilli exhibiting snapping-division
C. No fermentation of glucose, maltose or sucrose
D. Resistance to tellurite
E. Volutin granules

2.2 Which statement about *Corynebacterium diphtheriae* is FALSE?

F. Avirulent strains cannot stimulate protective immunity
G. Cells rich in interferon are resistant to toxin
H. Relationship between colonial type and virulence is variable
J. Toxoid immunization is indicated for Schick negative reactors
K. Virulent strains are invariably toxigenic.

2.3 For which organism is the demonstration of tumbling motility in cultures incubated at room temperature (about 22°C) a useful criterion for distinguishing it from organisms of similar morphology?

L. *Corynebacterium diphtheriae*
M. *Corynebacterium ulcerans*
N. *Erysipelothrix rhusiopathiae*
O. *Listeria monocytogenes*
P. *Propionibacterium acnes.*

Answers overleaf

2.1 C. This short fermentation set is used to separate pathogenic *Corynebacterium diphtheriae* from non-pathogenic *C. xerosis* and *C. hofmanni. C. diphtheriae* ferments glucose and maltose but not sucrose. The typical cell morphology is seen in methylene-blue or Albert's stained smears from cultures on Loeffler medium. Typically they are seen as delicate slightly curved beaded or barred rods with 2 or 3 metachromatic volutin granules per cell. Snapping, or incomplete division, is responsible for the characteristic "Chinese letters" arrangement of cells.
The characteristic colonial differences of the three epidemiological types is seen after culture on blood-tellurite medium, used for their selective isolation. Gravis strains form large grey-black daisy-head colonies, intermedius strains form minute flat black colonies and mitis strains form medium sized grey-black glistening colonies with a central elevation which gives them the appearance of a "poached egg".

2.2 J. A positive Schick test identifies individuals with inadequate existing immunity to diphtheria, as shown by the appearance of induration and redness at the site of injection of 0.5 ml purified standard diphtheria toxin 3—5 days after inoculation. A positive reaction demonstrates the individual has less than 0.03 units/ml of circulating antitoxin and is therefore likely to be susceptible to the disease. Heat inactivated toxin injected concomitantly into the opposite arm detects allergic false positive reactions.

2.3 O. Except for *Listeria monocytogenes*, the organisms listed are nonflagellate and nonmotile so that a positive motility test can at once exclude them, although a negative reaction does not exclude *L. monocytogenes.*

2.4 **Which is _NOT_ a property of *Bacillus anthracis?***

Q. Aerobic
R. Capsulated
S. Gram-positive bacillus
T. Motile
U. Spore bearer.

2.5 **Which organism has a distinctive colonial morphology described as "Medusa-head"?**

V. *Aspergillus fumigatus*
W. *Bacillus anthracis*
X. *Klebsiella pneumoniae*
Y. *Pseudomonas aeruginosa*
Z. Rough strains of *Streptococcus pneumoniae.*

2.6 **Which clostridium very rarely shows spores either on culture or from clinical material and yet is highly heat resistant, surviving several hours boiling?**

A. *Clostridium histolyticum*
B. *Clostridium oedematiens*
C. *Clostridium sporogenes*
D. *Clostridium tetani*
E. *Clostridium welchii.*

2.7 **What is _NOT_ a biological property of *Clostridium welchii?***

F. Alpha toxin neutralization is the basis of the Nagler reaction
G. Causal relationship to gas gangrene and food poisoning
H. Food poisoning strains are usually non-haemolytic
J. Spores of classical type A strains are killed by 5 min boiling
K. Types A — E are identified by capsular antigenic specificity.

Answers overleaf

2.4 T. *Bacillus anthracis* is non-flagellate and non-motile whereas all the other anthracoids (e.g. *B. cereus*, *B. subtilis* and *B. megaterium*) are flagellate and highly motile which is a useful, simple, differential character.

2.5 W. The anthrax bacillus grows as a compact colony with wavy outline and rough-looking surface. Examination with a low power lens shows it is composed of tangled masses of filaments and twisted cord-like projections of undivided and unseparated bacilli. This mass of interlaced filaments gives the impression of a tangled mass of curly hair, hence the terms, "Medusa-head" and "judge's wig".

2.6 E. *Clostridium welchii* typically fails to show obvious spores except in rare instances or when grown on special spore-forming media. Heat resistant strains responsible for most cases of *Cl. welchii* food poisoning may survive up to six hours boiling and yet show no obvious spores.

2.7 K. The five types of *Clostridium welchii* (A—E) are distinguished by the major lethal, minor lethal and non-lethal toxins they produce. All strains produce the major lethal toxin alpha toxin (lecithinase C) which is useful for identification of *Cl. welchii* by the Nagler reaction but does not help to separate the different types.

2.8 Which statement about *Clostridium welchii* is FALSE?

L. Always capsulate in tissues
M. Beta-haemolysis on horse blood agar due to theta toxin
N. Motile under anaerobic conditions
O. Saccharolytic, with abundant gas production
P. "Stormy-clot" fermentation of milk medium.

2.9 Which type of *Clostridium welchii* is specifically associated with enteritis necroticans in man?

Q. Type A
R. Type B
S. Type C
T. Type D
U. Type E.

2.10 Which test is the standard method used for identification of *Clostridium welchii?*

V. Alpha-toxin neutralization on serum or egg yolk plate
W. Biochemical reactions including "stormy-clot" reaction
X. Pathogenicity test with paired protected and unprotected animals
Y. Pattern of antibiogram
Z. Slide agglutination with type specific antiserum.

2.11 Which one of the following clostridia is flagellate, has round terminal spores and possesses proteolytic but no saccharolytic properties?

A. *Clostridium histolyticum*
B. *Clostridium septicum*
C. *Clostridium sporogenes*
D. *Clostridium tetani*
E. *Clostridium welchii.*

Answers overleaf

2.8 N. *Clostridium welchii* is rare among the clostridia in being nonflagellate and nonmotile. Clostridial motility is in fact tested under anaerobic conditions. The capsule is antiphagocytic and acts as an aggressin, aiding invasiveness. Theta toxin is prominent in type A strains but is usually absent from heat-resistant food-poisoning isolates which lack this haemolytic activity, a feature of value in their recognition.

2.9 S. Enteritis necroticans may follow consumption of food heavily contaminated with type C strains and must not be confused with the relatively mild, self-limiting toxic food poisoning caused by classical type A strains, or its specially heat-resistant variant. Type C strains produce an extensive though patchy necrosis of the small bowel, with bloody diarrhoea and possibly a fatal outcome.

2.10 V. This is the basis of the Nagler reaction for identification of *Clostridium welchii*. Opalescence produced by the action of lecithinase activity of alpha toxin is specifically neutralized by antitoxin. The only exception is given by *Cl. bifermentans* which has a serologically related lecithinase also specifically inhibited by *Cl. welchii* alpha antitoxin. The "stormy clot" reaction in milk is characteristic but not specific of *Cl. welchii*.

2.11 D. With the exception of *Clostridium tetani* all possess both proteolytic *and* saccharolytic properties in differing degrees or central or subterminal spores whereas *Clostridium tetani* has highly characteristic terminal "drumstick" spores. *Clostridium welchii* is exceptional in lacking flagella and by only very rarely showing spores unless specially cultivated on a sporing medium.

2.12 Which organism is a motile, Gram-positive bacillus that swarms on horse blood agar when incubated under optimum conditions?

F. *Bacillus anthracis*
G. *Clostridium tetani*
H. *Clostridium welchii*
J. *Klebsiella pneumoniae*
K. *Proteus mirabilis.*

Answers overleaf

2.12 G. *Clostridium tetani* alone fits the description. When cultivated anaerobically, it swarms as a fine, almost invisible veil over the surface of the plate. *P. mirabilis* is even better known as a motile swarmer, but of course it is Gram-negative. The other organisms are all non-flagellate and cannot swarm. *B. anthracis* is an aerobic Gram-positive spore bearer; *Cl. welchii* is an anaerobic Gram-positive spore bearer; *K. pneumoniae* is a Gram-negative bacillus.

3. Mycobacteria, Nocardia and Actinomyces

3.1 Which organism is "acid fast"?

A. *Bacteroides fragilis*
B. *Clostridium welchii*
C. *El Tor Vibrio*
D. *Lactobacillus acidophilus*
E. *Mycobacterium smegmatis*

3.2 Which group of organisms typically resist decolorization with an acidic solution of alcohol after being stained vigorously with a heated concentrated solution of a basic dye?

F. Chlamydiae
G. Mycobacteria
H. Mycoplasmas
J. Rickettsiae
K. Spirochaetes

3.3 Which mycobacterium may fail to grow from clinical specimens seeded onto Lowenstein-Jensen medium yet will grow readily on Dorset egg medium?

L.	*M. avium*	(avian bacillus)
M.	*M. bovis*	(bovine bacillus)
N.	*M. bovis var. BCG*	(attenuated bovine bacillus)
O.	*M. microti*	(vole bacillus)
P.	*M. tuberculosis*	(human bacillus)

Answers overleaf

3.1 E. Any mycobacterium is acid-fast although saprophytic species such as *M. smegmatis* are not fully acid and alcohol fast. Acid fastness is thought to be associated with their thick hydrophobic cell wall, rich in mycolic acid and complex waxes.
The cells resist simple staining procedures, but stain if treated vigorously with heated concentrated solutions. Once stained they resist decolorization with 20 per cent sulphuric acid whereas non-acid fast organisms lose the stain and take up a contrasting counter stain. The staining method employed is that of Ziehl-Neelsen.

3.2 G. Mycobacteria possess a thick cell wall rich in mycolic acid and other complex lipids. This makes the cell wall hydrophobic and impermeable to aqueous stains. When treated by Gram's method mycobacteria remain unstained. Ziehl-Neelsen introduced a satisfactory staining method based upon staining with a concentrated solution of carbol fuchsin for 5 minutes, aided by a short application of heat. When stained these organisms resist decolorization with strong (20 per cent) sulphuric or nitric acid, and for the pathogenic species, the addition of alcohol. The action of acid removes all red stain from the tissues and non-acid fast bacteria. Treatment with the counter stain produces a contrasting blue or green colour against which the red acid-fast bacilli are clearly seen. This stain is widely used in diagnostic microbiology to identify mycobacteria.

3.3 M. On primary isolation *M. bovis* may be partially or completely inhibited by glycerol in an L—J slope. This allows the human and bovine tubercle bacilli to be distinguished on culture, the human strain *M. tuberculosis* being stimulated by glycerol to give a luxuriant, eugonic growth whereas the bovine strain *M. bovis* may grow poorly as a sparse, dysgonic growth or perhaps not at all.
Dorset egg medium lacks glycerol and is not inhibitory for *M. bovis* or other mycobacteria. The BCG strain is well adapted to artificial culture.

3.4 Which laboratory examination will distinguish with certainty *Mycobacterium tuberculosis* from *Mycobacterium bovis?*

Q. Antibiotic sensitivity pattern
R. Histology of the clinical lesion
S. Pathogenicity for guinea-pig
T. Pathogenicity for rabbit
U. Ziehl-Neelsen film.

3.5 Niacin (nicotinic acid) production is a characteristic distinguishing property of which mycobacterium?

V. *Mycobacterium avium*
W. *Mycobacterium balnei*
X. *Mycobacterium bovis*
Y. *Mycobacterium kansasi*
Z. *Mycobacterium tuberculosis.*

3.6 What statement about atypical mycobacteria is FALSE?

A. Culture characteristics differ from those of *M. tuberculosis*
B. Human to human transmission rarely if ever occurs
C. Runyon classified them into Groups I—IV
D. They are opportunistic pathogens for man
E. They are more antibiotic sensitive than *M. tuberculosis.*

Answers overleaf

3.4 T. The essential biological difference between the human and bovine bacillus is their virulence for the rabbit. The bovine bacillus is highly virulent for rabbits by any route of inoculation and the human only slightly so. If the rabbit survives for six weeks and an autopsy shows lesions only in the lungs or kidneys the strain is *Mycobacterium tuberculosis.* With *Mycobacterium bovis* the rabbit may die before this time or if not it should be killed. At autopsy a progressive generalized tuberculosis will be found.

3.5 Z. Evidence of niacin production is a useful laboratory criterion for identification of *Mycobacterium tuberculosis. M. ulcerans* is the only other mycobacterium that produces niacin and this property helps to distinguish it from the "swimming pool disease" pathogen *M. balnei* (*M. marinum*) which, like *M. ulcerans* grows at 33°C. *M. ulcerans* only rarely infects man.

3.6 E. The atypical mycobacteria are noted for their high degree of drug resistance compared with most strains of *M. tuberculosis.* Human infections by them often require combined surgery and chemotherapy for cure.
On culture group I strains (e.g. *M. kansasi*) develop a bright yellow pigment only if exposed to light. Group II strains (*M. scrofulaceum*) produce an orange or yellow pigment independent of light. Group III strains (e.g. *M. intracellulare* — the Battey bacillus, related to the avian tubercle bacillus) produce little or no pigment. Thus they are called photochromogens, scotochromogens and nonchromogens respectively. Group IV strains (e.g. *M. fortuitum*) alone are rapid growers producing well developed colonies in less than 1 week whereas Groups I—III strains require 1—3 weeks or longer for growth.
Atypical mycobacteria have caused pulmonary, lymph node, bones and joints and disseminated infections in man, mostly due to *M. kansasi* and *M. intracellulare*, but the incidence in Britain is small.

3.7 Which statement about *Mycobacterium tuberculosis* is FALSE?

F. Desiccation in sputum results in loss of virulence
G. Exposure to UV light is lethal
H. Killed by heating to 62°C for 30 minutes
J. Relatively resistant to disinfectants
K. Unusually resistant to acids and alkalis.

3.8 Which pathogen does *NOT* satisfy "Koch's postulates"?

L. *Bacillus anthracis*
M. *Clostridium tetani*
N. *Corynebacterium diphtheriae*
O. *Mycobacterium leprae*
P. *Salmonella typhi.*

3.9 Which one of the following properties is possessed by nocardia but *NOT* by actinomyces and is thus a simple basis for their differentiation?

Q. Ability to grow in air
R. Branching morphology
S. Formation of hyphae-like filaments
T. Gram-positive stain reaction
U. Procaryotic nucleus.

3.10 Which statement about *Actinomyces israeli* is FALSE?

V. Appears as branching mycelium or diphtheroid forms
W. Classified as a bacterium, not a fungus
X. Colonies in tissue show peripheral radial "clubs"
Y. Man is the main reservoir of infection
Z. The organism is a strict aerobe.

Answers overleaf

3.7 F. Tubercle bacilli remain viable and virulent for weeks or months in dried sputum unless exposed to direct sunlight whose UV component has a strong bactericidal effect. Mycobacteria are highly sensitive to moist heat and the process of low temperature pasteurization at 62°C for 30 minutes is effective, although now generally replaced by "flash" pasteurization at a high temperature for a fraction of a minute. Pasteurization of milk has virtually eradicated bovine tuberculosis from Britain.

3.8 O. *Mycobacterium leprae* is an obligate human parasite which cannot be cultivated in artificial media. There is however no doubt that it is the causative organism of leprosy and it has been grown in experimentally immuno-deficient mice and hamsters.
The aetiological role of the other organisms to their respective diseases has been firmly established, fulfilling all of Koch's postulates, from extensive animal pathogenicity studies and also as a result of accidental laboratory infections in man.

3.9 Q. Although superficially similar in many respects nocardia are clearly aerobic and actinomyces anaerobic or micro-aerophilic. Both are opportunistic pathogens of man, producing abscesses and multiple sinuses with release of colonies described as "sulphur" granules in the pus. Actinomycosis mainly affects the neck and jaw, lungs, and the abdomen. Nocardiosis may affect the chest and brain.

3.10 Z. *Actinomyces israeli* is either anaerobic or microaerophilic and will not grow aerobically. Although it forms hyphae-like filaments it is a procaryotic cell with typical bacterial mucopeptide. The radial eosinophilic hyaline "clubs" formed in tissues result from host-parasite interactions and are not specific for actinomycotic infections.

3.11 Which statement about *Nocardia asteroides* is FALSE?

A. Acts as an opportunistic pathogen in man
B. Classified as a fungus
C. Forms rods and branching filaments
D. Produces aerial hyphae in culture
E. Stains Gram-positive and may be weakly acid-fast.

3.12 Which one of the following organisms, transmitted by rats is notable for its ability to grow in the bacillary phase or as naturally occurring cell-wall deficient L forms?

F. *Leptospira icterohaemorrhagiae*
G. *Rickettsia mooseri*
H. *Spirillum minus*
J. *Streptobacillus moniliformis*
K. *Yersinia (Pasteurella) pestis.*

Answers overleaf

3.11 B. Nocardia are typical procaryotic bacterial cells taxonomically closely related to the mycobacteria and less closely to the corynebacteria. Morphologically, the branching filaments resemble the actinomyces group although nocardia are strict aerobes whereas the actinomyces are anaerobic or microaerophilic. The colonies are often very similar to those of the mycobacteria, but differ by producing aerial hyphae which gives the colony a powdery appearance. Although the cells may stain "acid fast" they are less resistant to decolorization than the mycobacteria.
Nocardia asteroides produces opportunistic infections in man only when the host defences are severely compromised by disease or immunosuppressive therapy. The commonest form of infection is a brain abscess secondary to a pulmonary focus.

3.12 J. *Streptobacillus moniliformis* is present in the mouth of the wild rat and is transmitted to man by a rat bite (one form of rat bite fever) or by consumption of rat contaminated milk (Haverhill fever). The name is derived from its tendency in culture to grow as a mixture of chaining Gram-negative bacilli interspersed with a yeast-like mass of swollen filaments and globular forms, often filled with minute granules. This globular form, first described by Klieneberger as the symbiotic growth of a pleuropneumonia-like-organism (PPLO) was later recognized by Dienes as a cell wall deficient bacterial phase usually capable of reversion to the bacillary form. Thus the term L form was introduced after the Lister Institute where Klieneberger made her discovery.

4. Coliform Bacilli

4.1 Which organism ferments lactose but is invariably non-motile?

A. *Klebsiella pneumoniae*
B. *Proteus vulgaris*
C. *Pseudomonas aeruginosa*
D. *Salmonella typhi*
E. *Shigella flexneri.*

4.2 The IMVIC reaction is used in the examination of water supplies to distinguish intestinal strains from free-living strains of lactose fermenting coliforms. Which one produces indole, is methyl red test positive, Voges-Proskauer reaction negative and does not utilize citrate?

F. *Enterobacter aerogenes*
G. *Enterobacter freundii*
H. *Escherichia coli*
J. *Klebsiella pneumoniae*
K. *None of these.*

4.3 Which of the following will distinguish enteropathogenic *Escherichia coli* from harmless commensal strains?

L. Antibiogram
M. Production of cholera-like exotoxin
N. Possession of R-factors
O. H antigen analysis
P. O antigen analysis.

4.4 Which species includes enterotoxin-producing strains responsible for a cholera-like diarrhoeal disease?

Q. *Clostridium botulinum*
R. *Clostridium welchii*
S. *Escherichia coli*
T. *Staphylococcus aureus*
U. *Streptococcus faecalis.*

Answers overleaf

4.1 A. With the exception of *Klebsiella pneumoniae* all the other organisms fail to ferment lactose. *Shigella flexneri*, in common with all members of this genus, is non-motile, as is *Klebsiella pneumoniae.*

4.2 H. Intestinal strains (++——) are readily distinguished from free-living strains such as *Enterobacter aerogenes* (——++) but there are many intermediate reactions and some variability in the same species. The free-living strains are commonly present on plants and soil and their presence in a water supply is not indicative of faecal contamination. *Escherichia coli* on the other hand, is an intestinal commensal and its presence in a water supply is taken as evidence of faecal pollution.

4.3 P. Enteropathogenic serotypes are distinguished by "O" antigen analysis, initially by slide agglutination using standard antiserum. Serotypes 026, 055, 0111, 0119 and 0127 occur most frequently but some overlapping may exist with the more recently recognized enterotoxic and enteroinvasive serotypes which are also identified by somatic antigenic analysis.

4.4 S. Certain serotypes of *Escherichia coli* (O groups 6, 8, 15, 25, 27, 78, 148 and 159) produce a potent enterotoxin which by stimulating adenyl cyclase activity causes cyclic adenosine monophosphate (CAMP) to be formed with resultant gross increase in water and electrolyte secretion by the gut (cf. cholera toxin).

4.5 Which antigen of *Salmonella typhi* can completely inhibit agglutination of the organism by O antiserum?

V. Flagellar d antigen
W. O antigen factor IX
X. O antigen factor XII
Y. Pili antigens
Z. Vi-antigens.

4.6 Which species is a member of salmonella group D and has the antigenic formula 9, 12 (Vi):d:—?

A. *Salmonella enteritidis*
B. *Salmonella paratyphi A*
C. *Salmonella paratyphi B*
D. *Salmonella paratyphi C*
E. *Salmonella typhi*

4.7 Which statement about *Shigella flexneri* is FALSE?

F. Aerobic, nonmotile, piliated, Gram-negative bacillus
G. Definitive identification based on somatic antigen analysis
H. Initial screening based on fermentation reactions
J. Only 200 organisms required to initiate infection in man
K. R-factor transfer from *Escherichia coli* does not occur.

Answers overleaf

4.5 Z. On primary isolation some strains are inagglutinable in O antiserum due to an abundant outer layer of Vi antigen which prevents access of specific agglutinins. Vi antigen is a polymer of N-acetyl-aminohexuronic-acid and may enhance virulence by inhibiting phagocytosis.

4.6 E. *Salmonella enteritidis* has the antigenic formula 1, 9, 12: g, m: and is therefore also included in group D due to possession of the group antigen 9. It is readily differentiated from *S. typhi* by its possession of O antigen 1, phase 1 flagellar antigens g, m and lack of the distinctive (Vi) antigen. *S. paratyphi A, B and C* are respectively placed in groups A, B and C.
This question has been included merely to emphasize the importance of antigenic analysis in the classification and identification of the salmonellae. Detailed formulae need *not* be committed to memory although you should be *familiar with* their form and nomenclature.

4.7 K. Resistance transfer factor (RTF) was first described in relation to an outbreak of shigellosis in Japan in 1959 where genes coded for resistance to several antibiotics were transferred as plasmids by conjugation from a resistant *E. coli* donor to a previously sensitive *Sh. flexneri* recipient with the immediate acquisition of multiple antibiotic resistance.

4.8 Which salmonella NEVER exhibits flagella phase variation, the H antigen invariably in specific phase?

L. *Salmonella cholerae-suis*
M. *Salmonella paratyphi B*
N. *Salmonella paratyphi C*
O. *Salmonella typhi*
P. *Salmonella typhimurium.*

4.9 Which of the following organisms is an obligate aerobe?

Q. *Actinomyces israeli*
R. *Bacteroides fragilis*
S. *Clostridium tetani*
T. *Pseudomonas aeruginosa*
U. *Veillonella species.*

4.10 Which organism has been implicated in hospital acquired infection resulting from the use of contaminated disinfectant solutions?

V. *Haemophilus influenzae*
W. *Neisseria meningitidis*
X. *Pseudomonas aeruginosa*
Y. *Staphylococcus aureus*
Z. *Streptococcus pyogenes.*

Answers overleaf

4.8 O. *Salmonella typhi* is a monophasic, flagellated bacillus existing only in phase I. It does not agglutinate in polyvalent H non-specific phase 2 (group) antisera whereas it is very common to find that an unidentified clinical isolate or most other salmonellae will first agglutinate with poly H non-specific sera. It is of interest to note that *S. cholerae-suis* and *S. paratyphi C* have the same antigenic analysis for both O antigen and H antigen (formula 6, 7 : c : 1,5) except that *S. paratyphi C* also possesses the Vi antigen. *S. paratyphi B* (antigenic formula 1, 4, 5, 12 : b : 1, 2) and *S. typhimurium* (antigenic formula 1, 4, 5, 12 : i : 1, 2) can only be distinguished when they are in flagellar phase I by the possession of b and i antigens respectively.
S. typhi (antigenic formula 9, 12, Vi : d : —) is notable in containing the Vi antigen which is included with the O factors as it is in the outermost somatic antigen.
(This question is included merely to familiarize you with antigenic analysis. Specific formulae need not be committed to memory.)

4.9 T. *Pseudomonas aeruginosa* is an obligate aerobe, deriving its energy from oxidative reactions. It is a common laboratory practice to make use of its inability to grow under anaerobic conditions by including a plate inoculated with it in every anaerobic jar. The presence of growth after incubation indicates anaerobic conditions were not achieved.
The *Bacteroides*, *Clostridium* and *Veillonella* genera are all anaerobic; *Actinomyces* are anaerobic or microaerophilic.

4.10 X. *Pseudomonas aeruginosa* and even more commonly other pseudomonads are capable of survival and multiplication in disinfectant solutions, especially if the "in-use" solution has been slightly inactivated by contact with cork bungs, plastics and even hard water. Contaminated disinfectants have included Chloroxylenol, quaternary ammonium compounds and even diluted phenolic solutions.

4.11 Which statement about *Pseudomonas acruginosa* is FALSE?

A. Ferments lactose and hydrolyses urea
B. Motile Gram-negative bacillus
C. Opportunistic pathogen of man
D. Resistant to most antibiotics in common use
E. Responsible for infection of burns and skin grafts.

4.12 Which statement about *Pseudomonas aeruginosa* is FALSE?

F. Never acquires antibiotic resistance by conjugation
G. Non-lactose fermenting, motile, Gram-negative bacillus
H. Opportunistic human pathogen
J. Produces the pigments pyocyanin and fluorescin
K. Typed by bacteriophage and bacteriocin methods.

4.13 Which test is used to distinguish *Proteus mirabilis* from *Proteus vulgaris?*

L. Indole reaction
M. Motility
N. ONPG test
O. Phenyl alanine deaminase test
P. Urease activity.

Answers overleaf

4.11 A. *Ps. aeruginosa* fails to ferment lactose or produce urease. It is occasionally present in commensal flora of the gut. It does not initiate infection in the healthy subject but causes serious infections in children with fibrocystic disease, or after burns. Often associated with chronic varicose or pressure ulcers, chronic urinary tract infection and septic wounds.

4.12 F. Most isolates are sensitive to carbenicillin, gentamicin (and related aminoglycosides tobramycin and amikacin) and colistin (polymyxin E) which are the antibiotics of choice although there has been an increase in carbenicillin and gentamicin resistance due to R-factors. These extra-chromosomal plasmids contain resistance genes to one or more antibiotics plus genes which control the formation of the F pilus by which the R factors are transferred to an F^- recipient cell rapidly producing a resistant population. Conjugation occurs not only between *Ps. aeruginosa* but also between them and coliforms of the *Enterobacteriaceae* which increases the possibilities for transfer.
The other answers are true. Pyocyanin production gave the organism its popular name, the "blue pus bacillus". Phage and pyocine typing are used in epidemiological studies. Infections are usually associated with local or general immune deficiencies.

4.13 L. The only differential test given is the indole reaction, positive with *P. vulgaris* (usually ampicillin resistant) and negative with *P. mirabilis* (usually ampicillin sensitive). With the other tests, both organisms give the same results; P.A. deaminase and urease positive, ONPG negative and almost invariably motile, often showing the phenomenon of swarming.
Proteus species form part of the intestinal commensal flora and are a cause of urinary tract infections, especially *P. mirabilis.*

4.14 Which of these organisms characteristically exhibits the phenomenon of swarming?

Q. *Escherichia coli*
R. *Proteus vulgaris*
S. *Pseudomonas aeruginosa*
T. *Salmonella typhi*
U. *Shigella flexneri.*

4.15–17 For each organism on the left, select the most appropriate partner from the list on the right.

4.15 *Klebsiella pneumoniae*
4.16 *Proteus vulgaris*
4.17 *Shigella flexneri.*

V. Motile, capsulated, lactose fermenter
W. Motile, noncapsulated, non-lactose fermenter
X. Nonmotile, capsulated, lactose fermenter
Y. Nonmotile, noncapsulated, lactose fermenter
Z. Nonmotile, noncapsulated, non-lactose fermenter.

4.18 Which one of the following is an oxidase positive, non-motile, non-fermentative, Gram-negative diplobacillus associated with angular conjunctivitis?

A. *Corynebacterium diphtheriae*
B. *Haemophilus influenzae*
C. Koch-Weeks bacillus
D. *Moraxella lacunata*
E. *Pseudomonas aeruginosa.*

Answers overleaf

4.14 R. Most strains of *Proteus species* when grown on nutrient agar or horse blood agar do not form compact colonies but spread as a thin veil of surface growth in all directions from the point of inoculation. Morphologically the swarming cells are elongated and possess numerous peritrichous flagella. Should the strain be inoculated at two different points in the plate the swarming growth from both loci fuse imperceptibly but if this is repeated with inocula of two different strains the spreading edges are inhibited before union takes place and a small but distinct line of demarcation between the different strains is produced. The mutually inhibitory effect is known as the "Dienes phenomenon" and is of some value in the epidemiological typing of clinical isolates.

4.15–17 4.15 X: 4.16 W: 4.17 Z.
Klebsiella species are typified by heavy capsulation with resultant characteristic mucoid colonies which together with absence of motility readily allows differentiation from *E. coli*. Highly motile "swarmers" of *Proteus species* produce the typical spreading growth on non-inhibitory media. Positive reactions always prove more useful than negative reactions and so absence of motility in *Shigella species* is often used in the reverse situation where shigella are excluded by demonstration of motility in a suspicious clinical isolate.

4.18 D. *Moraxella lacunata* (Morax-Axenfeld bacillus) is said to be the causal agent of subacute conjunctivitis, affecting especially the corners of the eye, although it may simply be a secondary invader complicating a viral infection. The organism is typically a diplobacillus which forms characteristic colonies surrounded by small craters (lacunae, hence its name) when grown on serum medium.

4.19 **Which organism is commonly difficult to distinguish from neisseriae in Gram-films from clinical specimens?**

F. *Acinetobacter anitratus*
G. *Bacteroides fragilis*
H. *Brucella melitensis*
J. *Streptobacillus moniliformis*
K. *Vibrio parahaemolyticus.*

4.20 **Which organism characteristically produces a bright red pigment?**

L. *Actinomyces israeli*
M. *Lactobacillus acidophilus*
N. *Mycobacterium kansasi*
O. *Pseudomonas aeruginosa*
P. *Serratia marcescens.*

4.21 **Which organism is strictly limited to one portal of entry for the successful establishment of infection?**

Q. *Bacillus anthracis*
R. *Brucella abortus*
S. *Mycobacterium bovis*
T. *Vibrio cholerae*
U. *Yersinia pestis.*

4.22 **Which organism or group of organisms is highly susceptible to an acid pH and may even die when cultivated on media containing fermentable carbohydrate?**

V. *Alkalescens-dispar* group
W. *Alkaligenes faecalis*
X. *Lactobacillus acidophilus*
Y. *Mycoplasma pneumoniae*
Z. *Vibrio cholerae.*

Answers overleaf

4.19 F. This metabolically inactive organism appears predominantly as Gram-negative diplococci although bacillary forms may be seen in fluid cultures. The organism is an opportunistic pathogen found in the normal vagina and conjunctiva; in both sites it can be confused with *Neisseria gonorrhoeae.*

4.20 P. Aerial transmission of infection has been investigated by forming an aerosol of a distinctive organism of low virulence and then exposing settle plates for 1 hour in order to determine the sedimentation pattern and the effect of air currents on carriage of bacteria. *Serratia marcescens* has been frequently used for this purpose as its colonies develop a distinctive red pigment. (There is some risk in such procedures as the organism is an opportunistic pathogen).

4.21 T. *Vibrio cholerae* is not an invasive organism and cannot set up an infection under natural conditions by any route other than intestinal. All the other organisms are of high or moderate invasiveness and can invade the body by several portals of entry.

4.22 Z. *Vibrio cholerae* is rapidly killed by acid, and the bactericidal action of the gastric acid barrier is at least partly responsible for the low incidence (1—5 per cent) of disease in exposed susceptible individuals. Acid water supplies are rarely if ever implicated in outbreaks of cholera.

4.23 Which is NOT a property of *Vibrio cholerae?*

A. Dies rapidly at low pH
B. Grows well at high pH (9.0—9.5)
C. Produces indole and reduces nitrate
D. Synthesizes neuraminidase (receptor destroying enzyme)
E. Vaccine confers solid immunity of long duration.

4.24 Which statement about *Bacteroides fragilis* is FALSE?

F. Commonly sensitive to clindamycin and metronidazole
G. Frequent cause of post-operative cross infection
H. Obligate anaerobic Gram-negative bacillus
J. Severe infection may produce "endotoxic shock" syndrome
K. Usually resistant to penicillin and all aminoglycosides.

Answers overleaf

4.23 E. Killed cholera vaccine affords only partial protection for about 6 months when re-vaccination is required. Immunity following recovery from the disease is little better although it does stimulate specific IgA antibodies within the gut in addition to a humoral response. Studies on the mechanism of action of the enterotoxin may allow the development of a pharmacological blocking agent to supplement or even replace immunization.

4.24 G. *Bacteroides fragilis* is a prominent commensal of the gut and female lower genital tract. Together with aerobic coliforms it causes many mixed wound infections after gastro-intestinal and gynaecological surgery. Severity of infection varies from minor putrid suppuration to fatal septicaemia. Due to its sensitivity to oxygen and its ability to act only as an opportunistic pathogen, the possibility for its implication in cross infection is remote.

5. Small Gram-negative Bacilli

5.1 Which organism produces characteristic colonial iridescence when capsulated strains are cultivated on Levinthal's agar?

A. *Bacillus anthracis*
B. *Clostridium welchii*
C. *Haemophilus influenzae*
D. *Klebsiella pneumoniae*
E. *Streptococcus pneumoniae.*

5.2 Which one of the following properties of *Haemophilus influenzae* is ABSENT from non-capsulate strains grown *in vitro?*

F. Ability to synthesize endotoxin
G. Colonial iridescence
H. Requirement for X and V factors
J. Resistance to bacitracin in selective isolation media
K. Satellite phenomenon with staphylococci.

5.3 Which statement about *Haemophilus influenzae* is FALSE?

L. Capsulated strains are virulent
M. Pittman type b strains cause meningitis
N. Respiratory pathogenicity results from X-factor requirement
O. V-factor necessary for oxidation-reduction system
P. X-factor necessary for synthesis of respiratory enzymes.

Answers overleaf

5.1 C. Colonies of capsulated *Haemophilus influenzae* are distinctively iridescent when viewed in oblique light. Loss of capsule is associated with loss of iridescence.
The other organisms when virulent are all capsulated but are not iridescent.

5.2 G. Capsulated organisms alone exhibit the characteristic iridescence when viewed in obliquely transmitted light after growth for less than 12 hours on Levinthal's medium.

The other properties do not vary with capsulation.

5.3 N. Respiratory tract tropism is unrelated to X-factor requirement for biosynthesis of respiratory enzymes. Although non-capsulate strains may cause respiratory infections, most severe infections, with associated bacteraemia and meningitis, are due to capsular type b strains. The essential role of the capsule in virulence is shown by the high protective value of specific anticapsular antibody, whereas antibody to endotoxin has little effect.

5.4 Which is NOT a property of *Haemophilus influenzae*?

Q. Causes exacerbation of chronic bronchitis
R. Elicits antibodies which cross placental barrier
S. Exhibits the phenomenon of satellitism
T. Requires both X and V growth factors
U. Responsible for epidemics of meningitis.

5.5 Which statement about *Haemophilus ducreyi* is FALSE?

V. Causal agent of lymphogranuloma venereum
W. Gram-negative, non motile bacillus
X. Primary cultivation possible on enriched blood agar
Y. Requires X but not V factor for growth
Z. Transmitted by sexual intercourse.

Answers overleaf

5.4 U. *H. influenzae* type b is a prominent cause of meningitis in children under 3 years of age. Cases are sporadic and sibling contacts do not necessarily develop this disease, even when carriers of the organism.
The determinant of virulence is the capsule, and all serious cases are of type b. Resistance is conferred by specific anti-capsular antibody, and differences in susceptibility reflect the level of immunity acquired by colonization with different serotypes or homologous strains of low virulence. Thus children, after losing congenital immunity progressively develop natural active immunity which in the face of a heavy challenge with *H. influenzae* type b, or if the respiratory epithelium is already damaged by virus infection, might be insufficient to prevent severe infection and meningitis might develop. Contacts are unlikely to be equally susceptible although those who have recently outgrown their congenital immunity should be given prophylactic antibiotics.

5.5 V. *H. ducreyi* is responsible for the venereal disease called chancroid, or soft sore to distinguish it from the hard chancre of primary syphilis. After incubation period of 4 days an inflamed ulcer with a spreading irregular edge develops at the site of inoculation and there is associated inflammation of the draining lymphatics. There is usually tender enlargement of the draining lymph nodes which may suppurate and rupture spontaneously before healing with extensive scar formation. Surgical incision or needle aspiration may avoid this complication.
Diagnosis is established by isolation of the causative organism but rapid provisional diagnosis may be made by examination of films of exudate or aspirate. Lymphogranuloma venereum is caused by organisms of group A chlamydiae, similar to TRIC agents.

5.6 Which statement about *Bordetella pertussis* is FALSE?

A. Live attenuated strain used as vaccine
B. Most primary isolates are capsulated
C. Requires special culture medium (e.g. Bordet-Gengou)
D. Small, non-motile, Gram-negative bacillus
E. Transmitted principally by aerial droplets.

5.7 What is the biological basis for the LACK of any specific association of *Brucella abortus* infection with abortion in the human?

F. Absence of erythritol from the human placenta
G. Early production of maternal IgM antibody
H. Fetal immunological tolerance
J. Placental transfer of human maternal IgG antibody
K. Stimulation of human complement-fixing antibody.

5.8 Members of which bacterial genus may be differentiated by the selective inhibitory effect of the dyes thionine and basic fuchsin, when incorporated in high dilution in culture media?

L. *Brucella*
M. *Haemophilus*
N. *Mycobacterium*
O. *Neisseria*
P. *Vibrio.*

Answers overleaf

5.6 A. Prophylactic vaccines are prepared from killed, fully virulent (S) organisms and are offered to all normal children at 3 months either alone or incorporated in all the "triple vaccine" with tetanus and diphtheria toxoids. The vaccine has been the centre of controversy over its possible association with allergic encephalomyelitis which is a potentially disastrous but fortunately rare adverse reaction. Most strains isolated during the catarrhal stage of whooping cough are well capsulated and maximally virulent but isolates from the convalescent stage may be poorly capsulated or even rough and avirulent. This change is due to S → R variation. Bordet-Gengou agar contains peptone, potato extract, glycerol and 25 per cent fresh sheep's blood. It is thought that the albumin in the blood neutralizes toxic fatty acids in the medium and some media include charcoal for the same purpose. By incorporating penicillin in the medium contaminating nasopharyngeal commensals are inhibited, thus greatly increasing its selective activity and thereby improving recovery of *B. pertussis* from clinical specimens.

5.7 F. There is no evidence that cattle and human strains of *Brucella abortus* differ in virulence or growth requirements. The bovine placenta and fetal membranes contain erythritol which has a growth stimulating effect on *B. abortus* but this chemical is absent from the human decidua. Thus the difference in incidence of abortus is explained by different tissue tropism due to the presence or absence of erythritol.

5.8 L. *Brucella abortus* is inhibited by thionine, *Brucella suis* is inhibited by basic fuchsin and *Brucella melitensis* is inhibited by neither dye.

5.9 Which *Brucella species* does <u>*NOT*</u> require CO_2 when cultivated, does <u>*NOT*</u> produce H_2S and can grow in the presence of 1: 50,000 thionine and 1: 50,000 methyl violet dyes?

Q. *Brucella abortus*
R. *Brucella melitensis*
S. *Brucella suis*
T. Immunizing strain of *Brucella abortus* S 19
U. None of these.

5.10 Which one of the following pairs of micro-organisms CANNOT be distinguished from each other by direct agglutination without prior agglutinin absorption?

V. *Brucella abortus* and *Brucella melitensis*
W. Commensal and enteropathogenic strains of *Escherichia coli*
X. *Salmonella typhi* and *Salmonella paratyphi B*
Y. *Shigella flexneri* and *Shigella sonnei*
Z. Two different Griffiths' types of *Streptococcus pyogenes.*

Answers overleaf

5.9 R., *Brucella abortus* requires CO_2 produces H_2S and is inhibited only by thionine, whereas *Brucella suis* does not require CO_2 produces H_2S and is resistant to thionine but inhibited by methyl violet. *Brucella abortus* biotype 5 is similar to *Brucella melitensis* in some respects but infects cattle in Britain.

5.10 V. The brucella antigens responsible for serological specificity are of two different types, both present in all species at the same time but differing in their relative proportion, so that one antigen is dominant in *Brucella abortus* while the other predominates in *Brucella melitensis*. Antisera raised against one species will therefore cross-agglutinate the other. Prior absorption of the antiserum with just sufficient heterologous cells to completely remove all agglutinin to its dominant antigen, causes the anti-serum to lose its ability to agglutinate the heterologous species although it retains its ability to agglutinate the homologous species and is said to be monospecific.
The other pairs are all serologically distinct.

6. Spirochaetes

6.1 Which possesses an axial filament?

A. Any spirochaete
B. Branching forms of *Nocardia asteroides*
C. Filamentous forms of *Haemophilus influenzae*
D. Hfr strains of *Escherichia coli*
E. Pathogenic mycoplasmas.

6.2 Which organism stains Gram-negative with a very loose irregular spiral morphology and a tendency to colonize epithelial surfaces?

F. *Borrelia vincenti*
G. *Leptospira canicola*
H. *Spirillum minus*
J. *Treponema pallidum*
K. *Vibrio cholerae*

6.3 Which is <u>NOT</u> a property of *Leptospira icterohaemorrhagiae?*

L. Cannot be cultivated in non-living media
M. Demonstrable by dark ground and fluorescence microscopy
N. Induces type specific immunity in man
O. Shows active rotational but no translatory movement
P. Tight regular spiral with one or more secondary curves.

Answers overleaf

6.1 A. All spirochaetes possess an axial filament composed of several flagella-like filaments arising from each end of the organism, which winds around the protoplasmic cylinder of the cell body within the cell wall. The filaments are almost certainly the organelles of motility, composed of contractile flagellin protein.

6.2 F. Borrelia are the only spirochaetes with a loose spiral morphology. *Treponema pallidum* has rigid tight spirals and *Leptospira canicola* has the fine regular spiral typical of the leptospira. *Borrelia vincenti* is a normal oral commensal but may act as an opportunistic pathogen, especially in the mouth and throat.

6.3 L. Leptospira are readily cultivated aerobically at 30°C in peptone broth enriched with 10 per cent inactivated (heated at 56°C for 30 minutes) serum. They will also produce colonies on semi-solid nutrient agar supplemented by 10 per cent serum and haemoglobin.
All the other statements are true.

7. Rickettsiae, Chlamydiae and Mycoplasmas

7.1 Which of the following is an obligate intracellular parasite?

A. *Brucella abortus*
B. *Listeria monocytogenes*
C. *Mycobacterium tuberculosis*
D. *Neisseria meningitidis*
E. *Rickettsia prowazeki*

7.2 Which one of the following is *NOT* a property of the rickettsiae?

F. All species have bacterial-type cell walls and contain both RNA and DNA
G. Arthropods serve as natural host and transmission vector for most species
H. Their growth is enhanced by sulphonamides and inhibited by PABA
J. They are devoid of all enzymes active in intermediary metabolism
K. They have a predilection for the endothelial cells of small blood vessels.

Answers overleaf

7.1 E. Rickettsia form one of the two groups of obligate intracellular bacteria, the other being the chlamydiae. Viruses form the other group of intracellular parasites but differ fundamentally in structure and biology from the minute bacteria which reproduce by binary fission, possess both DNA and RNA, and have a classical bacterial ultra-structure, some even possessing flagella. Like viruses, they fail to grow except in living tissue cells and with the exception of *Coxiella burneti*, survive poorly outside the host cell. Rickettsiae were named in honour of H. T. Ricketts who first described them. *R. prowazeki* is the causal agent of classical, epidemic, louse-borne typhus fever.
Although the other organisms given as distractors are either frequently seen within cells or are truly intracellular parasites during infection, they can all be cultivated in a cell free medium.

7.2 J. Although rickettsiae generally lack the full complement of metabolic enzymes necessary for independent existence (*R. quintana* of trench fever can be cultivated on blood agar in 10 per cent CO_2), they do have an active metabolism including the enzymes of the Krebs cycle. The haemorrhagic manifestations characteristic of most infections are largely the result of widespread thrombosis caused by rickettsial growth in the endothelial cells of small blood vessels.
R. prowazeki (epidemic typhus) is unusual in having man as the natural reservoir and the louse active only in ensuring its transmission. *Coxiella burneti* (Q-fever) is even more exceptional in lacking an insect vector in relation to man who is infected by direct contact or by inhalation of contaminated dust. The inhibitory effect of PABA is unrelated to folic acid metabolism but results from substrate competition with P-hydroxy benzoic acid which interferes with the synthesis of co-factors for electron transport.

7.3 Which of the following retains its infectivity on exposure to desiccation in contaminated dust?

L. *Coxiella burneti*
M. *Mycoplasma pneumoniae*
N. *Neisseria meningitidis*
O. *Treponema pallidum*
P. TRIC agents.

7.4 Which property differentiates *Coxiella burneti*, the causal agent of Q-fever, from viruses?

Q. Airborne transmission to man
R. Arthropod vector transmission to animals
S. Obligate intracellular parasitism
T. Passage through 0.4 μm collodion filter
U. Possession of both DNA and RNA.

7.5 Which property can differentiate chlamydiae from viruses?

V. Ability to elicit delayed hypersensitivity
W. Filterable nature of infectious unit
X. Existence as obligate intracellular parasite
Y. Formation of inclusion bodies in infected cells
Z. Sensitivity to tetracyclines.

Answers overleaf

7.3 L. *Coxiella burneti* differs from rickettsiae by its resistance to drying and relative heat resistance. Thus it can be transmitted in farmyard dust from infected cattle and sheep and can be transmitted in contaminated milk submitted to the standard pasteurization procedure. Human infection is not arthropod-borne.

7.4 U. Viruses contain either DNA or RNA but never both. Other characters absent from viruses include bacillary morphology with murein and endotoxin in cell wall, binary fission, sensitivity to tetracycline and chloramphenicol and possession of enzyme systems of Krebs cycle, electron transport and protein synthesis.

7.5 Z. Chlamydiae are inhibited not only by the tetracyclines but also by chloramphenicol, streptomycin and with some strains, sulphonamides: viruses are uniformly resistant. Other differential features they possess include a bacterial-type cell wall with mucopeptide, both DNA and RNA, ribosomes, metabolic enzymes and multiplication by binary fission. The other properties listed are common to both chlamydiae and viruses.

7.6 Which one of the following is *NOT* a property of the mycoplasmas?

A. Convert to bacteria when subcultured on antibiotic-free medium
B. Colonies adhere and burrow into solid culture media
C. Identification is achieved by serological growth inhibition
D. Procaryotic cells lacking a rigid cell wall.
E. Susceptible to 56°C for 30 minutes.

7.7 Which one of the following is *NOT* a cell wall deficient form of bacteria?

F. L-form
G. Mycoplasma
H. Protoplast
J. Rough mutant
K. Spheroplast.

Answers overleaf

7.6 A. Although mycoplasmas are procaryotic cells lacking a rigid cell wall, they are not cell wall deficient bacteria and therefore can never revert to a bacterial form with a mucopeptide rigid wall as can an unstable L-form. They possess a classical cytoplasmic unit membrane and are very heat labile. When cultured on osmotically protected media, such as 20 per cent serum agar, growth is seen within 2—5 days. Identification is established by placing a range of different specific antiserum impregnated discs on a lawn of the test organism and after incubation the particular species is recognized by the inhibition of growth produced by the homologous antibody. Smears stained by Giemsa method clearly show the pleomorphic form, with filaments, cocci, globules and ring forms. The finer coccal forms may pass the normal $0.45\,\mu$ bacterial filter.

7.7 G. With the exception of mycoplasmas all the other organisms are undoubtedly of bacterial nature and represent varying degrees of loss of cell wall. A rough mutant has lost its outer layer of smooth antigenic determinants exposing the rough determinants of the basal core polysaccharides. A protoplast and an L-form have lost their entire cell wall and are left with only the cytoplasmic membrane which makes them somewhat fragile and osmotically sensitive. Spheroplasts are virtually restricted to the Gram-negatives where they represent cells that have lost their rigid mucopeptide layer but retain the outer coat of lipopolysaccharide somatic antigen. Mycoplasmas are very similar to permanent L-forms which lack the capacity to revert to the bacillary form, but differ in DNA base ratio, possess sterols in their cell membrane and are in certain cases, pathogenic to man and animals, which is still not proven for L-forms.

8. Fungi

8.1 **Which organism is classed as eucaryotic?**

A. *Candida albicans*
B. *Micropolyspora faeni*
C. *Nocardia asteroides*
D. *Streptobacillus moniliformis*
E. *Streptomyces madurae.*

8.2 **Which one of the following is *NOT* a property of *Candida albicans?***

F. Budding yeast forms contain mitochondria
G. Grows as a pseudomycelium in the tissues
H. Hyphal cells may enlarge to form chlamydospores
J. Inhibited by aminoglycoside antibiotics
K. Produces opportunistic infections in man.

8.3 **Which is the single MOST important feature which distinguishes *Candida albicans* from related organisms?**

L. Absence of capsules
M. Formation of chlamydospores
N. Formation of branching mycelium
O. Sensitivity to Nystatin
P. Yeast-like oval budding cells.

Answers overleaf

8.1 A. *Candida albicans* is a dimorphic fungus which grows as oval budding yeast cells and elongated hyphae. It has a nuclear membrane and mitochondria typical of eucaryotic cells. The other species listed are bacteria and therefore procaryotic.
Confusion over the correct taxonomic position of *S. madurae* has led to its inclusion in the *Streptomyces*, *Nocardia* and most recently its own specific genus *Actinomadura.*

8.2 J. *Candida albicans* is resistant to antibacterial antibiotics but is inhibited by nystatin, 5-fluorocytosine, amphotericin B and by topical gentian violet dye. Selective isolation is achieved by culture on Sabouraud's glucose agar medium, on which it grows luxuriantly, the surface growth as budding yeast forms and the submerged growth as mycelium. Distinctive chlamydospores are formed in cultures grown on corn meal agar. The organism forms part of the normal commensal flora of the upper respiratory tract, the gut, and female lower genital tract. It may become invasive in debilitated immunosuppressed individuals, or after prolonged treatment with broad spectrum antibiotics (e.g. tetracyclines). The best known clinical infection is thrush.

8.3 M. Chlamydospores are specialized thick walled spherical, 6—9 μ diameter resting phase yeast cells which differentiate at the tips of branching hyphae when the organism experiences nutritional deprivation, as when, cultured on the nutritionally poor medium "corn meal agar". Chlamydospores are a characteristic feature of *Candida albicans* whereas the other features listed are not distinguishing properties.

8.4 Which statement about dermatophytes is FALSE?

Q. Dermatophytes are skin-associated parasitic bacteria
R. Dermatophytoses are exemplified by athlete's foot and ringworm
S. They are classified according to their characteristic morphology
T. They infect only keratinized tissues
U. Sabouraud's glucose agar medium is used for their isolation.

Answers overleaf

8.4 Q. Dermatophytes are not bacteria but a group of parasitic fungi which cause superficial, non-invasive infections of keratinized tissues, namely skin, hair and nails. They are classified by their colonial and cellular morphology after growth on Sabouraud's glucose agar at 20°C into three groups, trichophyton, epidermophyton and microsporum. The superficial mycoses they cause include tinea pedis (athlete's foot) — *T. mentagrophytes* and *E. floccosum*) tinea corporis (ringworm of the body — *T. glabrosa*) and tinea capitis (ringworm of the scalp — *M. audouini* and *M. canis*). Most epidemic ringworm infections are transmitted by infected children either directly or indirectly through combs, hair clippers and chair backs whereas sporadic infections may come from infected animals (dog, cat and horse). Athlete's foot is spread by indirect contact through contamination of the floor in communal showers and swimming pools. Griseofulvin is used successfully in both athlete's foot and ringworm. When given orally it is incorporated into newly formed keratin where it has antifungal activity.

9. Viruses

9.1 Which one of the following RNA viruses is *NOT* included in the picornavirus group?

A. Coxsackieviruses
B. Echoviruses
C. Louping-ill virus
D. Poliovirus
E. Rhinovirus.

9.2 Which one of the following statements about coxsackieviruses is FALSE?

F. Diagnosis is established by serology and not by virus isolation
G. Groups A and B are distinguished by mouse pathogenicity tests
H. Mild or subclinical infections are common
J. They are classified as picornaviruses
K. They may be propagated in suckling mice.

9.3 Which one of the following is *NOT* a member of the paramyxovirus group?

L. Measles virus
M. Mumps virus
N. Parainfluenza virus
O. Respiratory syncytial virus
P. Rubella virus.

Answers overleaf

9.1 C. Although louping-ill virus is a small RNA virus, it is not a picornavirus (pico = small + RNA) but a member of the *Togaviridae* and the only vector-borne encephalitis virus indigenous to Britain where it causes disease in sheep, mainly in Scotland and the north of England, and rarely infects man. It is related to Russian spring—summer encephalitis virus.
Picornaviruses have 20—30 nm diam. RNA containing virions showing cubic symmetry and high resistance to ether. They multiply in the cytoplasm. They comprise 3 subgroups: enteroviruses (polio, coxsackie and echoviruses), rhinoviruses (human and animal types, including foot-and-mouth disease virus) and calciviruses (animal viruses).

9.2 F. Laboratory diagnosis is based upon isolation of the virus from faeces, throat washings or CSF depending upon the clinical syndrome. Serological methods are also used, particularly for group B viruses. Identification is by neutralization, complement fixation and haemagglutination-inhibition tests. Serological diagnosis involves demonstration of a significant rise in type-specific antibody in paired sera. Coxsackieviruses may be isolated and propagated in suckling mice but routine primary isolation of group B and certain group A strains is more successful in tissue culture.

9.3 P. Although rubella virus closely resembles the paramyxoviruses in morphology epidemiology and clinical features, it is of the alpha virus sub group of togaviruses. This RNA virion is large (150—250 nm) and spherical with an ether soluble envelope studded with short projections, enclosing the nucleocapsid. The virus may be propagated in a variety of cell cultures. Some lines (e.g. human amnion cells), show a distinctive cytopathic effect of diagnostic value whereas in others (e.g. green monkey kidney cells) this change is absent and indeed it protects these cells from the CPE of other viruses by the phenomenon of interference. This property is used as a means of detecting rubella virus in clinical specimens by first inoculating green monkey kidney cells with the specimen and 7—10 days later challenging with a culture of echovirus 11. The presence of rubella is indicated by inhibition of the CPE normally produced by echovirus.

9.4 Which statement about respiratory syncytial virus is FALSE?

Q. Belongs to paramyxovirus group of RNA viruses
R. Causes infantile bronchiolitis and bronchopneumonia
S. Maternally transmitted antibody does not prevent infection
T. Produces giant cells and syncytial cytopathic effect
U. Shares properties of haemagglutination, haemadsorption and haemolysin activity with other paramyxoviruses.

9.5 Which of the following respiratory viruses is of the DNA-containing class?

V. Adenovirus
W. Influenza virus
X. Picornavirus
Y. Respiratory syncytial virus
Z. Rhinovirus.

9.6 Which one of the following is an example of a DNA-containing tumour virus?

A. Avian leukosis virus
B. Mouse leukaemia virus
C. Mouse mammary tumour virus
D. Rous sarcoma virus
E. Simian vacuolating virus (SV40).

Answers overleaf

9.4 U. Syncytial virus differs significantly from other paramyxoviruses (e.g. mumps, measles, parainfluenza and Newcastle disease viruses) by lacking the group's characteristic properties of haemagglutination, haemoadsorption by infected tissue culture cells and haemolysin activity. All paramyxoviruses have a similar morphology, the virion possessing a helical capsid, containing the genome of single-stranded RNA, within a lipid-rich, ether sensitive envelope, covered with regularly spaced short projections.

9.5 V. The adenoviruses are medium sized viruses with double stranded DNA, lack envelopes and are ether resistant. There are at least 33 human serotypes of which only a small number are important as pathogens.

9.6 E. Simian vacuolating virus or SV40 occurs commonly in uninoculated rhesus monkey kidney cell cultures in which it produces no cytopathic effect, but if introduced into kidney cell cultures from green monkeys it causes characteristic vacuolation of the cytoplasm. The virus is known to produce experimental tumours in hamsters and will cause transformation of cell cultures derived not only from animals but also from man. Millions of recipients of both inactivated and live virus vaccines have inadvertently been given SV40 as a contaminant and the virus has subsequently been isolated for up to 5 weeks after ingestion of live polio vaccine. SV40-like virus is associated with progressive multifocal leukoencephalopathy in man and has also been isolated from renal transplant patients on immunosuppressives. The closely related papilloma (human wart) virus, polyoma virus and SV40 virus together comprise the DNA papova group of tumour viruses, the name being derived from the first two letters of each name (PA = papilloma; PO = polyoma; VA = vacuolating). The other viruses listed are all RNA tumour viruses of the oncornavirus group (oncogenic + RNA).

Part Three

MICROBIAL DISEASES AND CLINICAL MICROBIOLOGY

1. Transmission of Infective Disease

1.1 The term epidemic is used to denote a disease outbreak which affects a large proportion of a human population. What is the animal equivalent?

A. Endemic
B. Epizootic
C. Pandemic
D. Pestilence
E. Sporadic.

1.2 Which statement about droplet nuclei is FALSE?

F. Formed by evaporation of exhalation droplets
G. Rarely contain more than a single infectious particle
H. Responsible for spread of exogenous chest infections
J. Under normal ward conditions settle within 30 minutes
K. Usually penetrate the lower respiratory tract.

1.3 In which disease does the human carrier play an essential role in the evolution of an epidemic?

L. Bubonic plague
M. Meningococcal meningitis
N. Q-fever
O. Tuberculosis
P. Weil's disease.

Answers overleaf

1.1 B. Epizootic is used for diseases of animals which are highly infectious or contagious, rapidly spreading and widely diffused. *Endemic* is applied to diseases which are prevalent in a specific region but which are neither epidemic nor sporadic. *Pandemic* is used to describe a widespread epidemic. *Pestilence* is a general term for any virulent contagious or infectious epidemic disease. *Sporadic* is used in contradistinction to epidemic, meaning that it occurs intermittently in isolated cases here and there.

1.2 J. Droplet nuclei sediment very slowly in still air (0.04 ft/min) and are readily resuspended by upward air currents of only 1.0 ft/min which is greatly exceeded by air movement in any naturally ventilated room (upwards of 25 ft/min). Thus they remain suspended indefinitely until removed by ventilation or if trapped on a surface (oiled surface). Fortunately the contained organisms rapidly die or lose their virulence, partly due to desiccation and partly to the lethal effect of UV component of sunlight which may be supplemented by the use of shortwave UV lamps.

1.3 M. *Neisseria meningitidis* is transmitted by aerial droplets from healthy carriers before cases of meningitis occur. Normally 2—10 per cent of healthy individuals harbour the pathogen, but in the pre-epidemic period the carriage rate increases progressively. Clinical cases appear when carriage reaches 30 per cent and this may rise to 90 per cent at the height of an epidemic.
Of the other diseases bubonic plague alone is epidemic, but depends for its spread on an infected, flea-infested rat population, and temporary human carriage is unimportant. Q-fever and Weil's disease are strictly zoonoses. Tuberculosis is spread by man almost exclusively by active cases of open pulmonary disease, and although quiescent cases are common reactivation of a primary infection must again develop into open clinical disease before transmission is likely.

1.4 Which organism is *NOT* characteristically transmitted by healthy carriers?

Q. *Bordetella pertussis*
R. *Neisseria meningitidis*
S. *Salmonella typhi*
T. *Staphylococcus aureus*
U. *Streptococcus pyogenes.*

1.5 Which one of the following is LEAST likely to produce a chronic carrier state or to play the role of a harmless commensal?

V. *Actinomyces israeli*
W. *Bordetella pertussis*
X. *Haemophilus influenzae*
Y. *Haemophilus parainfluenzae*
Z. *Pseudomonas aeruginosa.*

1.6 In which one of the following diseases is man the natural host and sole reservoir of infection?

A. Bacillary dysentery
B. Botulism
C. Clostridial food poisoning
D. Salmonella food poisoning
E. Staphylococcal food poisoning.

Answers overleaf

1.4 Q. Patients harbour *B. pertussis* during the catarrhal, paroxysmal and for about two weeks during the whooping stage. Transmission is by aerial droplets disseminated during coughing with an attack rate of 70 per cent in unvaccinated children and up to 90 per cent in house contacts.

1.5 W. *B. pertussis* is spread only by infected persons, especially during the early catarrhal phase. The chronic carrier state is not recognized but it may be maintained in a population by mild cases of the disease between epidemics every 2 to 5 years. Controversy over the risk of vaccine encephalopathy has adversely affected the vaccination rate ("triple vaccine" includes *B. pertussis* killed cells with diphtheria and tetanus toxoids) so that serious epidemics may result from an increase in susceptible children to a critical level.

The other organisms listed all regularly occur as normal commensals of man, acting as opportunistic pathogens.

1.6 A. Bacillary dysentery is a human infection caused by members of the genus Shigella. The organisms are transmitted by persons with clinical disease or with mild or subclinical infection. Convalescent carriers are common whereas chronic carriers are rare. Strains responsible for disease in Britain are *Sh. sonnei* (mild), *Sh. flexneri* (intermediate) and very rarely *Sh. dysenteriae* (severe).
The other diseases all have sources other than man. *Cl. botulinum* exists both as a free-living saprophyte in soil and as an intestinal parasite in animals. *Cl. welchii* is widely distributed in soil and as a gut commensal in man and animals. Food poisoning strains of salmonella exist mainly as animal parasites whereas the enteric fever group is limited to man. Although staphylococcal food poisoning is generally of human origin, *Staph. aureus*, phage type 42D is a common cause of bovine mastitis and milk contaminated with it has been the cause of several outbreaks of food poisoning.

1.7 In which disease is communicability between humans negligible?

F. Leprosy
G. Pertussis
H. Q-fever
J. Salmonellosis
K. Trachoma.

1.8 Which is the source of MOST infections caused by *Vibrio parahaemolyticus?*

L. Crabs, shrimps and shellfish
M. Cream cakes and pastries
N. Drinking water
O. Fried rice from a Chinese restaurant
P. Raw milk and dairy produce.

1.9 Which disease is *NOT* a zoonosis?

Q. Anthrax
R. Bubonic plague
S. Leprosy
T. Leptospiral jaundice
U. Listeriosis.

1.10 Which one of the following human pathogens is spread by animal hosts?

V. *Bordetella pertussis*
W. *Listeria monocytogenes*
X. *Neisseria meningitidis*
Y. *Treponema pallidum*
Z. *Vibrio cholerae.*

Answers overleaf

1.7 H. Human infection with *Coxiella burneti* results almost exclusively by drinking contaminated raw milk or inhaling infective dust from animal bedding. The spread of leprosy, pertussis and trachoma is restricted to human hosts and the food poisoning species of salmonella are often associated with human sources, although their main natural reservoir is usually an animal species.

1.8 L. Shellfish and crustaceans contaminated with *Vibrio parahaemolyticus* have caused numerous outbreaks of food poisoning in Japan and America, and is now increasingly recognized as a cause of acute gastroenteritis in Britain. It should be remembered that various *Salmonella species* may also be transmitted by contaminated seafood.

1.9 S. Leprosy is an entirely human disease with no natural reservoir of infection in any animal species. All of the other diseases are transmitted to man directly or indirectly from animal hosts in which the pathogen has a natural stage of development, normally with the production of disease.

1.10 W. *Listeria monocytogenes* has a wide range of animal hosts. Human infection may result from direct or indirect contact with infected animals. For the others, man is the only natural host although animals may be infected experimentally with *T. pallidum* maintained for use in immobilization tests.

1.11 Which organism is known to have important natural reservoirs of infection other than man?

A. *Bordetella pertussis*
B. *Listeria monocytogenes*
C. *Neisseria meningitidis*
D. *Treponema pallidum*
E. *Vibrio cholerae.*

1.12 Which is the MOST unlikely vehicle of infection in a case of human anthrax in Great Britain?

F. Bones and bone meal
G. Butcher meat
H. Carcasses
J. Hides and skins
K. Wool and hair.

1.13 Which animal is *NOT* implicated in transmission of anthrax to man?

L. Cow
M. Fowl
N. Goat
O. Horse
P. Sheep.

1.14 The importation of goat hair yarn is prohibited because of the risk of contamination with which of the following?

Q. Anthrax spores
R. *Brucella melitensis*
S. Cysts of *Entamoeba histolytica*
T. Smallpox virus
U. *Vibrio cholerae.*

Answers overleaf

1.11 B. *Listeria monocytogenes* infects a broad spectrum of wild and domestic animals exhibiting a variety of disease patterns in different host species. Rabbits and man alone show the characteristic monocytosis. Human infection occurs from direct or indirect contact with infected animals, or after ingestion of raw milk and dairy products. Infections in pregnancy cause metritis and abortion, stillbirth or neonatal septicaemia. Meningitis and septicaemia also occur in the adult but inapparent genital carriage is more common.

1.12 G. Anthrax in a cattle beast is usually suspected or diagnosed in life and notified under the Diseases of Animals Act which governs the handling and disposal of the carcass, thus preventing the possible sale of contaminated meat. Most human infections result from contact with contaminated imported wool and hair, hides and skins, and bones and bone meal. It is therefore a particular hazard in woollen mills, tanneries, bone meal factories and among dock workers. The disease is rare in Britain, with less than 10 cases occurring annually with perhaps one death. Propylactic vaccination, with annual boosters, is offered to workers at special risk.

1.13 M. Fowls are naturally immune to anthrax, possibly related to their relatively high body temperature, as the resistance can be overcome experimentally by simply lowering their temperature. Pasteur infected fowls by immersing their legs in water at a temperature of 20°C and another worker achieved similar success by lowering the birds' temperatures by antipyrin.
All the other animals are susceptible to anthrax and they or their products may be a source of human infection.

1.14 Q. Goat hair yarn is commonly contaminated with anthrax spores and cannot be satisfactorily disinfected without sustaining damage unacceptable to manufacturers.

1.15 Which non-sporebearing organism, transmitted by contact with fish, poultry and swine, causes a spreading, red, non-suppurative lesion of the skin in man?

V. *Erysipelothrix rhusiopathiae*
W. Lysogenized strains of *Streptococcus pyogenes*
X. *Staphylococcus aureus*, phage type 71
Y. *Streptococcus pyogenes* of Griffiths type 4
Z. Virulent strains of *Bacillus anthracis.*

1.16 Which is *NOT* an arthropod-borne infection of man?

A. Bubonic plague
B. Epidemic typhus fever
C. European relapsing fever
D. Q-fever
E. Scrub typhus.

1.17 Which is *NOT* implicated in the spread of epidemic typhus?

F. Cases of Brill-Zinsser disease
G. Cases of Classical or European typhus fever
H. Chronic human carriers of *Rickettsia prowazeki*
J. *Pediculus corporis*
K. Wild rats.

1.18 Which method provides the BEST assessment of the level of bacterial contamination of operating theatre air?

L. Determine the bacterial content of the air supply filter
M. Determine the post-operative sepsis rate
N. Expose settle plates in theatre
O. Measure nasal *Staph. aureus* carriage rate in theatre staff
P. Use slit-sampler apparatus in theatre.

Answers overleaf

1.15 V. *Ery. rhusiopathiae* causes the cutaneous infection erysipeloid acquired by handling infected animals in fish markets, abbattoirs and bone meal factories. The others produce lesions of the skin distinguishable either clinically, epidemiologically or bacteriologically.

1.16 D. Q-fever (*Cox. burneti*) is transmitted by milk and dust and is vector-borne only for animals. Bubonic plague (*Y. pestis*) is transmitted by the rat flea (*Xenopsylla cheopis*) although pneumonic plague is spread from man to man by respiratory secretions. Epidemic typhus fever (*R. prowazeki*) and European relapsing fever (*Borr. recurrentis*) are transmitted by the human body louse (*Pediculus corporis*) whereas scrub typhus (*R. tsutsugamushi*) is spread by mites (*Trombicula species*).

1.17 K. Humans with epidemic typhus (synonym Classical or European) and chronic carriers of *R. prowazeki* with recrudescent typhus (Brill-Zinsser disease) constitute the only reservoir of infection in contrast to most rickettsial diseases which have animal reservoirs. The disease is transmitted by the human body louse, which contaminates its bite with infective faeces. Delousing, with residual insecticide, controls the disease.

1.18 P. Although settle plates provide a simple method for assessing aerial contamination results are inaccurate. The slit sampler technique is more accurate as it examines much larger volumes of air. Measured volumes of air drawn through a narrow slit impinge on a slowly rotating Petri culture plate, depositing the airborne bacteria. The degree of contamination is assessed by counting the colonies after incubation. By knowing the sampling rate and speed of rotation the airborne count can be monitored throughout the run, identifying times of particular risk which may be correlated with specific activities.

1.19 What is the probable source of post-operative wound infection, caused by mixed coliforms and *Bacteroides fragilis*, in a patient who had full mechanical and prophylactic antibiotic bowel preparation prior to colectomy?

Q. Member of surgical team
R. Operating theatre aerial contamination
S. Patient's commensal gut flora
T. Surgical instruments or sutures
U. Ward dressings procedure.

1.20 What category of operation is specifically associated with significant post-operative sepsis due to *Staphylococcus albus* and diphtheroid organisms?

V. Abdomino-perineal resection of rectum
W. Dental extraction
X. Partial gastrectomy
Y. Total hip prosthetic replacement
Z. Tracheostomy.

1.21 Which procedure is the MOST practical and certain method of preventing brucellosis in the general population of Great Britain at this time?

A. Brucellin-testing of dairy herds
B. Isolation of human cases
C. Pasteurization of all milk
D. Slaughter of infected cattle
E. Vaccination of calves.

Answers overleaf

1.19 S. Such infections are almost invariably of endogenous origin due to leakage from the opened gut or later from the anastomosis. Prophylactic antibiotics do not reduce the incidence of such infections and should be avoided with the possible exception of metronidazole which is thought to reduce the incidence of bacteroides sepsis. Typing of the infecting organisms often confirms their identity with strains present in the patient's gut commensal flora.

1.20 Y. Replacement of a diseased head and neck of femur by a plastic or stainless-steel prosthesis, thermally-cemented into the upper end of the femoral shaft, is a lengthy and traumatic operation which involves the deliberate implantation of a massive foreign body into the tissues. Even with the most scrupulous aseptic surgical technique and laminar flow protection small numbers of skin commensals may gain entry to the tissues and contaminate the prosthesis, possibly penetrating the porous cement. The prosthesis thus greatly reduces the number and virulence of organisms necessary to establish infection, by protecting them from phagocytosis. This permits low virulence species, such as *Staph. albus* and diphtheroids, which regularly contaminate clean surgical wounds, to evade phagocytic elimination and multiply with the production of disease and ultimate failure of the operation.

1.21 C. Although all these methods lessen the risk by removing the potential source, pasteurization remains the one simple and sure preventive measure at present. The ultimate aim would be to eliminate infection from domestic animals but the vast expense and the limited degree of immediate success of this policy leaves pasteurization the mainstay of present control. It should be noted that isolation of human cases is unnecessary as human to human transmission is very rare.

1.22 Which preventive measure has been MOST effective in reducing the incidence of intestinal tuberculosis caused by *Mycobacterium bovis?*

F. BCG immunization of population
G. Introduction of attested herds
H. Meat inspection and condemning of tuberculous carcasses
J. Pasteurization of milk
K. X-ray screening of dairy workers.

1.23 Which organism transmitted by the bite of sandflies, is the causal agent of two distinct clinical diseases, Oroya fever and Verruga peruana?

L. *Bartonella bacilliformis*
M. *Bdellovibrio bacteriovorus*
N. *Rickettsia rickettsi*
O. Sandfly fever virus
P. *Streptobacillus moniliformis.*

Answers overleaf

1.22 J. The greatest reduction in bovine tuberculosis followed the widespread pasteurization of milk at a time when elimination of infected cattle from a herd was based solely on the result of clinical examination combined with bacteriological examination of the milk.

1.23 L. *Bartonella bacilliformis* is of the order *Rickettsiales*, differing from classical rickettsiae by motility and ability to grow on non-living artificial media. It shows a characteristic tendency to multiply within or on the surface of erythrocytes, appearing in blood films as a minute pleomorphic coccobacillus. *B. bacteriovorus* is a minute ectoparasite which grows only in contact or within the host bacterium probably due to lack of enzyme systems essential for independent growth. *R. rickettsi* is the causal organism of tick-borne typhus, Rocky Mountain spotted fever. The sandfly fever virus is transmitted by the sandfly *Phlebotomus papatasi* hence the name "phlebotomus fever". *S. moniliformis* is the agent of Haverhill fever and one type of rat-bite fever. (cf Soduku, caused by *Spirillum minus*).

2. Bacterial and Mycotic Infections

2.1 **Which organism is mainly responsible for boils, carbuncles, impetigo, paronychia, breast-abscess and osteitis?**

A. *Candida albicans*
B. *Propionibacterium acnes*
C. *Proteus vulgaris*
D. *Staphylococcus aureus*
E. *Streptococcus pyogenes.*

2.2 **Which organism is the MOST probable cause of neonatal septicaemic illness with destructive metastatic abscesses in bone?**

F. *Actinomadura madurae*
G. *Actinomyces israeli*
H. *Listeria monocytogenes*
J. *Mycobacterium tuberculosis*
K. *Staphylococcus aureus.*

2.3 **Which is LEAST likely to be found in an untreated severe infection with *Staphylococcus aureus?***

L. Anti-staphylolysin in the serum
M. Initial suppurative focus
N. Leucocidin induced leucopenia
O. Metastatic abscess formation
P. Septicaemia.

Answers overleaf

2.1 D. *Staphylococcus aureus* is the most common pathogen carried on the skin and is therefore frequently associated with superficial infections, wound infections and subcutaneous abscesses. It also quite often breaches the surface defences and may invade the bloodstream and be deposited in the growing end of a young bone, causing acute osteitis. Subacute or chronic paronychia caused by *Staph. aureus* is clinically very similar to that caused by *Candida albicans. P. acnes* may be recovered from boils but is not considered to be a primary pathogen. *Strep. pyogenes* also causes superficial pyogenic infections and abscesses but is much less common than *Staph aureus* and tends to produce spreading cellulitis, or erysipelas, rather than a boil. It is frequently associated with impetigo.

2.2 K. *Staphylococcus aureus* neonatal osteitis is a serious and sometimes fulminating disease which can produce devastating destructive lesions often destroying the growing end of the bone. In mothers infected with, or carrying, *Staph. aureus* the baby is rapidly colonized; the umbilical stump, skin, upper respiratory tract and external auditory canal are the common sites from which invasive infection may follow.

2.3 N. Such infections, commonly seen in the pre-antibiotic era, originate by invasion of the circulation from a suppurative focus which results in a distant metastatic abscess. Secondary and more serious invasion of the bloodstream may then occur as a septicaemia, with widespread metastatic visceral and skeletal abscesses, terminating in death. Even in such an overwhelming infection leucocytosis typical of staphylococcal infections persists at a high level with the leucocidin merely contributing to pus formation. Normal uninfected individuals regularly have a low serum anti-staphylolysin titre and this would probably be raised due to the initial suppurative focus even if terminally the infection overwhelmed the immune system.

2.4 Place the following events in their proper order and indicate which would come FOURTH?

Q. Osteitis
R. Pyaemic abscesses
S. Septic skin lesion
T. Septicaemia
U. Transient intermittent bacteraemia.

2.5 Place the following events in their proper order and indicate which would come FOURTH?

V. Acute rheumatic fever
W. Mitral stenosis
X. *Streptococcus pyogenes* throat infection
Y. *Streptococcus viridans* bacteraemia
Z. Subacute bacterial endocarditis.

2.6 Which does *NOT* contribute to the incidence of subacute bacterial endocarditis?

A. Acute rheumatic fever
B. Dental surgery
C. Patent ductus arteriosus
D. Transient bacteraemia
E. Vaccination with living attenuated organisms.

Answers overleaf

2.4 T. Minor staphylococcal lesions of the skin not infrequently produce a transient bacteraemia which is normally rapidly cleared by the reticulo-endothelial system phagocytes in spleen and liver. Occasionally in young or adolescent boys and even in the newborn, *Staph. aureus* becomes localized in the metaphysis or growing end of a long bone with the production of osteitis or osteomyelitis. Constant heavy invasion of the bloodstream from this acute focus produces the clinical syndrome of septicaemia which in turn may cause widespread visceral and skeletal abscesses characteristic of pyaemia. Although the outcome of infection has been radically altered by the introduction of antibiotics active even against penicillinase-producing *Staph. aureus*, osteitis is still a serious disease liable to produce deformity of the growing bone, become chronic or even prove fatal.

2.5 Y. Throat infection with *Streptococcus pyogenes* precedes by about 1 to 5 weeks the development of rheumatic fever, an immunologically mediated disease characterized by fever, polyarthritis and pancarditis which may produce permanent damage to the heart valves. The mitral valve is most often affected with fibrosis of the cusps producing mitral stenosis and incompetence. Subsequently, a transient bacteraemia may occur with *Strep. viridans* possibly after tooth extraction. The abnormal heart valve is prone to colonization by circulating organisms which when established leads to the formation of thrombus, the source of emboli which are characteristic of SBE.

2.6 E. Any factor which causes damage to the endothelium lining of the heart, heart valves or major vessels or which promotes transient bacteraemia increases the chance of acquiring subacute bacterial endocarditis. Vaccination with live vaccine does not give rise to SBE. In susceptible individuals undergoing dental surgery, prophylactic penicillin is indicated.

2.7 In which disease do the symptoms appear generally in the first week after contracting the infection?

F. Brucellosis
G. Diphtheria
H. Leprosy
J. Syphilis
K. Tuberculosis.

2.8 Which is the LEAST likely result of a severe infection with *Corynebacterium diphtheriae?*

L. Asphyxia
M. Cervical adenitis
N. Myocardial degeneration
O. Peripheral neuritis
P. Septicaemia.

2.9 Which one of the following is associated with acute ulcerative gingivostomatitis (trench mouth)?

Q. *Candida albicans*
R. *Corynebacterium diphtheriae*
S. *Corynebacterium ulcerans*
T. *Treponema pallidum*
U. *Vincent's organisms*

Answers overleaf

2.7 G. The average incubation period for diphtheria is from 2—5 days, occasionally longer. Leprosy has by far the longest incubation period of 1—2 years. Symptoms of brucellosis usually develop in 2—4 weeks, syphilis about 1 month and tuberculosis within 4—6 weeks. These are only the average incubation periods and individual cases may show considerable variations depending upon the weight of infecting dose, virulence of organism, portal of entry, general health of the subject and level of specific resistance.

2.8 P. The virulence of *Corynebacterium diphtheriae* depends essentially upon two properties. Firstly its ability to breach the mucosal barrier to establish infection. Secondly, its ability to produce a potent exotoxin which is the main mechanism of pathogenicity, the disease being almost entirely a toxaemia. Even highly virulent strains remain closely associated with the portal of entry showing no tendency to invade the circulation to cause septicaemia.

Asphyxia results from respiratory obstruction by the pseudomembrane and cervical adenitis results from reaction of the local draining lymph nodes. Myocarditis and peripheral neuritis result from the direct effect of toxin carried to them in the systemic circulation.

2.9 U. This is a variety of Vincent's infection (cf. Vincent's angina affecting the pharynx) which occurs as a secondary complication of damage to the mucous membrane by virus infection (*Herpes simplex*), nutritional deficiency (often B-complex) or mechanical injury (carious teeth and ill-fitting dentures). This opportunistic infection results from a symbiotic overgrowth of commensal spirochaetes (*Borrelia vincenti*) and cigar-shaped fusiform bacilli (anaerobic Gram-negative bacilli of the *Bacteroides — Fusiformis* group). It is amenable to penicillin therapy but may recur if the initial cause of mucosal damage persists. All of the listed organisms may cause oral lesions but only Vincent's organisms are specifically associated with ulcerative gingivitis.

2.10 Which organism is the commonest cause of acute epiglottitis?

- **V.** *Haemophilus influenzae*
- **W.** *Listeria monocytogenes*
- **X.** *Mycobacterium tuberculosis*
- **Y.** *Neisseria meningitidis*
- **Z.** *Streptococcus pneumoniae.*

2.11 In which infection caused by *Haemophilus influenzae* is the acute inflammatory response potentially very dangerous?

- A. Acute epiglottitis
- B. Acute exacerbation of chronic bronchitis
- C. Acute sinusitis
- D. Bronchopneumonia
- E. Meningitis.

2.12 Which one is *NOT* the result of infection with *Haemophilus influenzae?*

- F. Acute bronchitis and bronchopneumonia
- G. Acute epiglottitis and laryngotracheitis
- H. Maxillary sinusitis
- J. Primary atypical pneumonia
- K. Purulent meningitis.

2.13 Which respiratory tract pathogen establishes infection by adhering specifically to the ciliated epithelium of the respiratory tract?

- L. *Bordetella pertussis*
- M. *Haemophilus influenzae*
- N. *Klebsiella pneumoniae*
- O. *Mycoplasma pneumoniae*
- P. *Streptococcus pneumoniae.*

Answers overleaf

2.10 V. Epiglottitis, usually caused by *Haemophilus influenzae* type b, occurs mainly in children and may prove fatal unless the respiratory obstruction is relieved by emergency tracheostomy. The age incidence is very similar to that to *H. influenzae* meningitis.

2.11 A. The vascular component of the acute inflammatory response to infection often results in escape of plasma proteins from engorged and permeable capillaries with the production of local oedema. While this is generally beneficial in aiding localization of the infection, it may present a serious threat to life in acute epiglottitis, particularly in children, when the oedematous epiglottis can obstruct the airway necessitating emergency tracheostomy.

2.12 J. Although *Haemophilus influenzae* is a common commensal of the upper respiratory tract where it acts as an opportunistic pathogen, it does not cause the clinical syndrome of primary atypical pneumonia. This is caused by a heterogenous group of organisms including influenza, and parainfluenza viruses, adenovirus, respiratory syncytial virus, group B chlamydiae (psittacosis), *Coxiella burneti* (Q-fever) and pre-eminently *Mycoplasma pneumoniae.*

2.13 L. During the early catarrhal stage of whooping cough large numbers of infective respiratory droplets are emitted during mild coughing. When inhaled by a susceptible individual, lacking adequate protective immunity, *Bordetella pertussis* shows a particular predilection to adhere to the cilia and surface of the epithelium of the trachea and bronchi. This allows the pathogen to establish itself locally where it multiplies rapidly and causes much local damage by the action of its endotoxin. Although the local tissues are infiltrated with neutrophil polymorphs in response to tissue breakdown, the blood shows an absolute lymphocytosis of up to 30,000/mm^3. *Bordetella pertussis* shows no tendency to invade the bloodstream and rarely persists even locally after the convalescent phase.

2.14 Which one is the LEAST likely site for a primary tuberculous focus?

Q. Lower lobe of lung
R. Meninges
S. Skin
T. Terminal ileum
U. Tonsil.

2.15 Place the following events in a logical sequence and indicate which should come FOURTH?

V. Caseation of hilar lymph node
W. Miliary tuberculosis
X. Primary (Ghon) focus in lung
Y. Rupture into pulmonary vein
Z. Tuberculous meningitis.

2.16 Which statement about pulmonary tuberculosis is FALSE?

A. BCG vaccination affords only partial protection
B. Especially prevalent in persons with silicosis
C. Primary isolation of *M. tuberculosis* takes from 3–6 weeks
D. Prolonged combined chemotherapy is essential
E. Three successive negative cultures exclude tuberculosis.

Answers overleaf

2.14 R. Tuberculous meningitis is probably always secondary to a focus elsewhere in the body. The other sites are all possible portals of entry of tubercle bacilli. Primary infection of the ileum caused by *M. bovis* is now rare since the introduction of pasteurization of milk although endogenous post-primary gut lesions still occur from swallowed tuberculous sputum. Primary lesions in the skin are an occupational hazard of laboratory workers, butchers, postmortem room attendants and laundry workers.

2.15 W. The sequence of events is most likely to be X, V, Y, W and Z. Entrance of large numbers of bacilli into the systemic circulation results in acute dissemination of the disease with the development of minute foci of infection throughout the body. Tuberculous meningitis is often associated with miliary tuberculosis.

2.16 E. Tubercle bacilli often occur in large numbers in the sputum in cases of active pulmonary tuberculosis. Laboratory diagnosis is based upon demonstration of "acid fast" bacilli in a Ziehl-Neelsen or auramine stained film and cultivation for *M. tuberculosis* on Lowenstein-Jensen slopes, usually positive within 3—6 weeks. It is common practice to examine 3 consecutive morning specimens of sputum as a standard screening for open tuberculosis. Excretion of tubercle bacilli may be intermittent and in some cases active tuberculosis may be present in the absence of excretion of "acid fast" bacilli in sufficient numbers to ensure detection by microscopic examination of a Z.N. film.

2.17 Which one of the following is *NOT* usually associated with lepromatous leprosy?

F. Abundant "acid fast" bacilli in the lesions
G. Continuous bacteraemia
H. Positive lepromin skin test
J. Progressive disease with poor prognosis
K. Slow development of symmetric skin lesions.

2.18 Which statement about actinomycosis is FALSE?

L. Cervico-facial, pulmonary and abdominal disease occurs
M. Colonies released in pus are called "sulphur granules"
N. Definitive diagnosis depends upon isolation of causative organism
O. Limited epidemics occur in farming communities
P. Typically results in multiple draining sinuses.

2.19 Which is the commonest cause of uncomplicated urinary tract infections?

Q. *Escherichia coli*
R. *Proteus mirabilis*
S. *Pseudomonas aeruginosa*
T. *Staphylococcus albus*
U. *Streptococcus faecalis.*

Answers overleaf

2.17 H. The progressive or lepromatous phase occurs when the individual shows high susceptibility to the disease. The lepromin test (analogous to the Mantoux reaction) is usually negative whereas in the tuberculoid phase the lepromin test is always positive. All the other findings are characteristic of lepromatous leprosy.

2.18 O. Actinomycosis results from endogenous infection secondary to local trauma such as dental extraction or appendicectomy. The disease is sporadic with no evidence of transfer from animals or man to man. The greater prevalence of the disease in farming communities is probably the result of greater carriage of the organisms in rural communities where it is primarily a disease of farm animals. (A presumptive diagnosis may be made on typical histology).

2.19 Q. *Escherichia coli* is by far the commonest cause of urinary tract infections in patients with no history of recurring incidents. The infection is endogenous; the gut commensal is either mechanically introduced into the urinary tract or becomes invasive, entering the circulation and passing to the glomerular tuft. A limited number of serotypes are principally implicated, (O serotype 1, 2, 4, 6, 7, 50 and 75), but this may merely reflect the commensal spectrum carried in the normal gut.
The other organisms are more often isolated from cases with some local or general predisposition to infection, such as obstruction or stasis due to physical abnormality of the urinary tract or adjacent tissues.

2.20 Which statement about acute urinary tract infection in pregnancy is FALSE?

V. *Escherichia coli* is usually responsible
W. Generally amenable to correct antibiotic therapy
X. Microscopy permits a provisional rapid diagnosis
Y. Pregnancy is purely coincidental and does not predispose to infection
Z. Urine colony count generally exceeds 10^5 per ml.

2.21 Which one of the following conditions is *NOT* a clinical manifestation of infection with *Staphylococcus aureus?*

A. Acute bacterial endocarditis
B. Acute osteomyelitis
C. Food poisoning
D. Scalded-skin syndrome
E. Septicaemia.

2.22 Which one of the following aetiological agents of bacterial food poisoning has an incubation period of 2—6 hours?

F. *Clostridium botulinum*
G. *Clostridium welchii*
H. *Salmonella enteritidis*
J. *Salmonella typhimurium*
K. *Staphylococcus aureus.*

Answers overleaf

2.20 Y. There is a complex relationship between pregnancy and urinary tract infection which involves the obvious anatomical and hormonal changes as well as the risk of infection being introduced during physical examination and catheterization. Pressure on the bladder and ureters from the enlarging uterus may cause partial obstruction with stasis so that transient low number contaminants may be allowed time to multiply due to incomplete voiding of urine, despite frequency of micturition. This may be complicated by local, hormonal and enzyme effects on smooth muscle and connective tissue tonus, causing among other things dilation of the ureters. Pregnancy is also a period of suppressed immunological reactivity (e.g. allograft rejection delayed) and it seems probable that blood and tissue clearing mechanisms may be partially inhibited which may encourage the establishment of infection.

2.21 C. Staphylococcal food poisoning is *not an infection* but an intestinal *intoxication* which results from consumption of preformed enterotoxin produced in contaminated unrefrigerated food. The source of contamination is usually a food handler who is a healthy carrier, often with a skin condition such as eczema, or a minor staphylococcal lesion. Staphylococcal food poisoning must not be confused with the serious infection, post-operative enterocolitis, which results from extensive infection of the intestine with *Staph. aureus* after loss of the normal commensal flora from overenergetic use of broad spectrum antibiotics.

2.22 K. Source is usually a food handler carrying or infected by an enterotoxin producing strain of *Staph. aureus*, 90 per cent belonging to bacteriophage group III. The vehicle of transmission is often pastries or cold meats unrefrigerated for over 4 hours. Salmonella and clostridial food poisoning generally have an incubation period of from 12—24 hours, or longer.

2.23 Which one of the following staphylococcal diseases is LEAST likely to be of long duration?

L. Carbuncle
M. Food poisoning
N. Osteitis
O. Lung abscess
P. Septic arthritis.

2.24 Which organism causes serious outbreaks of gastroenteritis in neonates and infants?

Q. *Alcaligenes faecalis*
R. *Enterobacter cloacae*
S. *Escherichia coli*
T. *Pseudomonas aeruginosa*
U. *Streptococcus faecalis.*

2.25 Which situation is LEAST likely to result in diarrhoeal disease?

V. *Escherichia coli* (enterotoxin-producing serotype) in tourist
W. *Escherichia coli* (enteropathogenic serotype) in nurse
X. *Salmonella typhimurium* in geriatric patient
Y. *Shigella flexneri* in school child
Z. *Vibrio cholerae* in Indian peasant.

2.26 What is the usual incubation period of salmonella food poisoning?

A. 2–6 hours
B. 12–36 hours
C. 2 weeks
D. 4 weeks
E. More than 1 month.

Answers overleaf

2.23 M. The symptoms of staphylococcal food poisoning appear within 2—6 hours, persist for several hours and subside completely within 24 hours leaving no residual morbidity. Even where large numbers of *Staph. aureus* are consumed with the contaminated food, infection is only very rarely established, due to a combination of high gastric acidity, intact surface defences, competition by the commensal flora and the inhibitory effect of the intestinal anaerobic environment.

2.24 S. Enteropathogenic serotypes of *Escherichia coli* (e.g. O serotypes 55, 111, 119 and 128) are distinct from the gut commensal opportunistic pathogens and the specific enterotoxin producing serotypes. Enteropathogenic strains can penetrate gut epithelium producing acute and even fatal gastroenteritis in babies and infants. Symptoms include vomiting, diarrhoea and in severe cases profound dehydration. Recent admissions to a nursery with mild or inapparent infection are the usual sources of serious outbreaks with possibly some deaths.

2.25 W. *Escherichia coli* gastroenteritis decreases in incidence throughout the pre-school period due to active acquired immunity from increasing exposure to antigens. Adults may be carriers and serve as a source of infection but do not develop diarrhoeal disease.

2.26 B. Symptoms of nausea, vomiting, diarrhoea and fever begin 12—36 hours after the incriminated meal, the time varying with the number of ingested organisms and the general health of the victim. Early in the disease faecal excretion is high but later becomes intermittent before finally ceasing.

2.27 Which salmonella serotype is MOST frequently isolated in Great Britain from cases of salmonella food poisoning?

F. *S. agona*
G. *S. anatum*
H. *S. enteritidis*
J. *S. heidelberg*
K. *S. typhimurium.*

2.28 What method offers the best chance of showing that clinical isolates of *Salmonella typhimurium* originated from a common source?

L. Antigenic analysis
M. Bacteriophage typing
N. Determination of antibiogram
O. Determination of biotype
P. Evaluation of DNA homology.

2.29 Which statement about typhoid fever is FALSE?

Q. Chloramphenicol therapy increases relapse rate
R. Chronic carriers often have gallbladder infection
S. Leucopenia characteristically occurs
T. Man is the sole natural reservoir
U. TAB vaccination results in solid immunity.

2.30 Which salmonella is MOST likely to cause osteitis of the tibia or vertebrae?

V. *S. agona*
W. *S. anatum*
X. *S. enteritidis*
Y. *S. typhi*
Z. *S. typhimurium.*

Answers overleaf

2.27 K. All the given serotypes are commonly isolated from outbreaks of food poisoning but *S. typhimurium* isolates generally equal or exceed the combined total of all other serotypes. *S. typhimurium* is also frequently isolated from veterinary salmonellosis, particularly from cattle.

2.28 M. Bacteriophage typing is a rapid and reliable method with high discriminating value and therefore suited to epidemiological studies. Antigenic analysis is limited by the predominance of a few strains responsible for most outbreaks. Biotype and antibiograms have little discriminatory value and DNA homology is concerned with genetic relatedness at generic level.

2.29 U. TAB vaccine confers only a minimal protection in as much as it raises the number of organisms required for a minimum infective dose. Recovery from infection confers a much higher and more prolonged resistance but even this is not complete. Chloramphenicol inhibits agglutinin formation and this may account for its paradoxical effect on the relapse rate. Cholecystectomy ends the carrier state in about 85 per cent of carriers which emphasizes the important epidemiological role of gallbladder infection. Intermittent faecal excretion of *S. typhi* may continue after clinical cure — *convalescent excretors* for a short period, *temporary excretors* for up to one year and *chronic carriers* when over one year. Excretors with no history of clinical infection are termed *symptomless excretors.*

2.30 Y. During the initial septicaemic phase of typhoid fever, organisms may lodge in the bones and remain unnoticed until some later date. Widal reaction should be tested when investigating bony lesions especially if there is a history of enteric fever or infection with *S. cholerae-suis.*

2.31 Which salmonella characteristically causes septicaemia with no enteric involvement?

A. *S. cholerae-suis*
B. *S. paratyphi A*
C. *S. paratyphi B*
D. *S. typhi*
E. *S. typhimurium.*

2.32 Which one of the following is *NOT* an infection?

F. Athlete's foot
G. Botulism
H. Erysipeloid
J. Impetigo
K. Madura foot.

2.33 Which organism is specifically associated with incidents of food poisoning after eating Chinese fried rice?

L. *Bacillus cereus*
M. *Clostridium welchii*
N. *Salmonella typhimurium*
O. *Staphylococcus aureus*
P. *Vibrio parahaemolyticus.*

Answers overleaf

2.31 A. Salmonella infections are of three clinical types — gastro-enteritis or food poisoning, enteric fever and the septicaemias. *S. cholerae-suis* typically produces a septicaemia without gut lesions, although metastatic visceral and skeletal abscesses or foci of infection are quite common. *S. typhimurium* is almost exclusively associated with gastroenteritis being the major cause of salmonella food poisoning. *S. typhi and S. paratyphi A* and *B* are the classical causal organisms of enteric fever.

2.32 G. Botulism is a serious form of food poisoning caused by ingestion of food containing preformed exotoxins of the anaerobic sporebearer *Cl. botulinum*. The source is inadequately sterilized preserved meats and vegetables, especially after home canning and bottling. Spores surviving the food processing, later multiply with production of a potent neurotoxin. Symptoms appear 12—24 hours after the meal. Progressive cranial nerve paralysis develops with death occurring from respiratory paralysis or cardiac arrest. Administration of antitoxin may be life saving.
Only 12 cases of the disease in Britain had been recorded until summer 1978 when an outbreak occurred in Birmingham involving 4 elderly people, two of whom died. The source of the toxin was thought to have been an imperfectly sealed can of North American salmon. Botulism is more common in USA, probably due to popularity of home bottling of vegetables.

2.33 L. Incidents of acute food poisoning following the consumption of Chinese fried rice are now quite common.
B. cereus is the most frequently implicated species. Holding cooked, fried rice at only a moderate temperature permits multiplication of sporebearers which survived the initial cooking temperature.

2.34 Which organism may give rise to a benign, self-limiting form of acute mesenteric adenitis which may be easily mistaken for acute appendicitis in children?

Q. *Bacillus cereus*
R. Enteropathogenic *Escherichia coli*
S. *Shigella sonnei*
T. *Vibrio parahaemolyticus*
U. *Yersinia pseudotuberculosis.*

2.35 Which organism is particularly associated with the Waterhouse-Friderichsen syndrome?

V. *Clostridium tetani*
W. *Corynebacterium diphtheriae*
X. *Mycobacterium tuberculosis*
Y. *Neisseria meningitidis*
Z. *Shigella sonnei.*

2.36 Neonatal meningitis is MOST commonly caused by which one of the following organisms?

A. *Escherichia coli*
B. *Haemophilus influenzae*
C. *Listeria monocytogenes*
D. *Mycobacterium tuberculosis*
E. *Neisseria meningitidis.*

Answers overleaf

2.34 U. The finding of a normal appendix and ileum, together with enlarged inflamed mesenteric lymph nodes, at laparotomy for possible acute appendicitis, suggests *Y. pseudotuberculosis* infection. Diagnosis is confirmed by isolation of the causal organism from a lymph node biopsy, or by detecting specific agglutinins against *Y. pseudotuberculosis*, (90 per cent of cases are due to type 1). Thus the infection may be confirmed either pre-operatively in suspected cases or after laparotomy where culture was omitted or gave negative results.

2.35 Y. The W.—F. syndrome is profound peripheral circulatory collapse with acute adrenal insufficiency as a result of extensive bilateral haemorrhagic necrosis of the adrenals. This may appear as a grave complication of any very acute septicaemia but is particularly associated with fulminant meningococcaemia in children. Meningococcaemia is characterized by thrombo-embolic lesions due to metastatic foci of infection and local and generalized Shwartzman reactions to endotoxin, which produce an acute depletion of coagulation factors with tendency to haemorrhage.
Although the W.—F. syndrome is not associated with diphtheria, adrenal haemorrhage is prominent in experimental diphtheria in rabbits and small haemorrhages occur in human disease.

2.36 A. *E. coli* and other enteric bacilli are the major cause of meningitis in the newborn. Group B streptococci are also important at this period. The prevalence of *E. coli* may be related to early colonization of the bowel in bottle fed babies, at a time when the immune system is immature. A more direct relationship is seen in *E. coli* neonatal meningitis in babies with congenital abnormalities that effect the CNS and its membranes, such as myelomeningocoele and spina bifida, where anatomical defects expose the vulnerable tissues to direct invasion by coliforms derived from the individual's own commensal flora or the hospital resident flora.

2.37 Which age group is particularly prone to *Haemophilus influenzae* meningitis?

F. Under 3 months
G. 3 months—3 years
H. 3 years—12 years
J. 12 years—25 years
K. Over 25 years.

2.38 What organism causes a benign lymphocytic meningitis often with associated fever, nephritis and conjunctivitis?

L. *Brucella melitensis*
M. *Cryptococcus neoformans*
N. *Haemophilus aegyptius*
O. *Leptospira canicola*
P. *Listeria monocytogenes.*

2.39 Which organism causes a mild form of meningitis in which the CSF shows a lymphocytic pleocytosis, increase in protein and normal or slightly decreased glucose and chloride levels?

Q. *Haemophilus influenzae*
R. *Leptospira canicola*
S. *Neisseria meningitidis*
T. *Pseudomonas aeruginosa*
U. *Streptococcus pneumoniae.*

Answers overleaf

2.37 G. Babies are born with protective congenital immunity which progressively diminishes over the first three months of life. By the age of three years natural active immunity to the invasive capsulated type b strains has been gradually acquired by repeated exposure to related organisms or specific strains of low virulence. Thus an inverse relationship exists between the incidence of meningitis and the level of type-specific antibody in the blood. Cases which occur in the adult may be related to either intense respiratory infection with *H. influenzae* capable of invading the bloodstream, or to the presence of immune deficiency.

2.38 O. Leptospirosis is the cause of about 5 per cent of all cases of aseptic meningitis. The association of meningitis, nephritis and conjunctival suffusion, in a patient with a febrile illness, would be very suggestive of this diagnosis.

2.39 R. Leptospiral infection commonly shows two phases. The more virulent strains, which cause serious infection, accentuate the early bacteraemic phase symptoms of fever, rigors, severe headache, episcleral injection and possibly jaundice and purpura. The CSF can be shown by dark ground microscopy to contain leptospira and these may be recovered on culture. The second "immune phase" follows after a brief remission, with the recurrence of meningeal symptoms and CSF changes suggestive of benign aseptic meningitis. The CSF shows a moderate lymphocytic pleocytosis with raised protein and normal or slightly decreased glucose and chloride levels. Organisms are usually not demonstrable but serial antibody studies show a significant rising titre. (Even if you did not know this you should have recognized that all the other organisms produce frank pyogenic meningitis.)

2.40 Which organism is associated with an infection of the fetus or newborn called granulomatosis infantiseptica?

V. *Candida albicans*
W. *Erysipelothrix rhusiopathiae*
X. *Listeria monocytogenes*
Y. *Mycobacterium tuberculosis*
Z. *Yersinia pseudotuberculosis.*

2.41 Which statement about brucellosis is FALSE?

A. Blood culture is valueless in acute infection
B. Chronic disease is usually diagnosed serologically
C. Intermittent fever is a common symptom
D. Usually transmitted by raw milk
E. Very rarely associated with abortion in the human.

2.42 Which organism is commonly associated with septic dog and cat bites?

F. *Haemophilus haemoglobinophilus*
G. *Leptospira canicola*
H. *Pasteurella multocida*
J. *Toxocara canis*
K. *Toxoplasma gondi*

2.43 *Woolsorter's disease* is caused by which one of the following?

L. *Bacillus anthracis*
M. *Clostridium tetani*
N. Louping-ill virus
O. Orf virus
P. None of these.

Answers overleaf

2.40 X. *L. monocytogenes* is transmitted from the vagina in intra-uterine infection of the fetus. The other distractors have been chosen because of their obvious relationship to granulomatous infection except for *E. rhusiopathiae*, chosen because it may be confused with *L. monocytogenes. L. monocytogenes* is a not uncommon cause of meningitis in children and occasionally in adults.

2.41 A. Blood culture is positive in the first 2—3 weeks before the immune response eliminates all extracellular brucellae, leaving those surviving within macrophages to continue the infection. Blood cultures, inoculated into liver-infusion or trypticase-soy broth, should be incubated at 37°C in 10 per cent CO_2 for up to 8 weeks. Serological diagnosis may be made by agglutination reactions after the second week. This is the basis of diagnosis in chronic infection. PUO is the usual presentation. Abortion is rare as the human placenta lacks the growth factor erythritol, present in the bovine placenta.

2.42 H. All are carried by dogs or cats but only *Past. multocida* is commonly recovered from infected bites. A mucoid strain, not derived from animals, occasionally causes human respiratory tract infection. *Past. multocida* is sensitive to penicillin, the drug of choice.

2.43 L. Pulmonary anthrax is a severe respiratory infection caused by the inhalation of dust contaminated with anthrax spores. The disease is an occupational hazard in woollen mills and factories where imported wool and hair are sorted and cleansed.
The other related diseases are: M. tetanus; N. louping-ill of sheep and encephalitis in man; O. orf or contagious pustular dermatitis of sheep and rarely man.

2.44 Which is the most likely causal organism of an epidemic of acute superficial conjunctivitis in a school?

Q. Group A chlamydiae
R. *Haemophilus aegyptius* (Koch-Weeks' bacillus)
S. *Moraxella lacunata*
T. *Neisseria gonorrhoeae*
U. *Streptococcus pneumoniae.*

2.45 What is the usual incubation period of gonorrhoea?

V. 24 hours
W. 4 days
X. 2 weeks
Y. 4 weeks
Z. 3 months.

2.46 Which statement about gonorrhoea is FALSE?

A. Communicability negligible in asymptomatic females
B. Contraceptive pills encourage vaginal colonization
C. No effective immunity to reinfection
D. Prolonged carriage occurs in both women and men
E. Transmission is by direct and indirect contact.

Answers overleaf

2.44 R. *Haemophilus aegyptius* is responsible for about 40 per cent of acute conjunctivitis and for the vast majority of outbreaks of contagious conjunctivitis. It was first observed by Robert Koch (in 1883) in an investigation of conjunctivitis in Alexandria when he was working on the aetiology of cholera. Weeks (in 1887) later cultivated similar organisms from cases of acute contagious conjunctivitis. Since then the organism has been thoroughly characterized and shown to be closely related to but not identical with *Haemophilus influenzae* and has been officially named *H. aegyptius*. Its causal relationship to an epidemic form of acute superficial conjunctivitis, so-called "pink eye", has been established. Trachoma, produced by group A chlamydiae, is a much more chronic disease.

2.45 W. Gonorrhoea in the male, typically has an abrupt onset, about 4 days after exposure, with symptoms of a profuse mucoid discharge, rapidly becoming purulent, inflammation of the urinary meatus and painful, burning sensation on micturition. Microscopic examination of the discharge shows numerous neutrophils, some containing large numbers of Gram-negative diplococci, which permits a presumptive diagnosis to be made, but this should be confirmed by culture and identification of the pathogen. Female cases are less frequently seen in the early stage of the disease, possibly because the onset is less abrupt but it seems probable that there is a comparable short incubation period.

2.46 A. Infection usually results from venereal contact with an asymptomatic female chronic carrier. Contraceptive pills promote gonorrhoea by encouraging sexual promiscuity, affording no physical barrier to infection unlike condoms or by raising the normally acid pH of the vagina to neutrality thus allowing colonization and growth of this delicate pathogen. This view may be somewhat controversial.

2.47 Which is LEAST likely to result as a complication of untreated gonorrhoea?

F. Endocarditis
G. Inguinal bubo
H. Polyarthritis
J. Prostatitis
K. Uveitis.

2.48 Which organism is associated with sterility in women and retention of urine in men?

L. *Brucella abortus*
M. *Candida albicans*
N. *Haemophilus vaginalis*
O. *Neisseria gonorrhoeae*
P. *Trichomonas vaginalis.*

2.49 Which clinical condition is *NOT* caused by TRIC-agents?

Q. Inclusion conjunctivitis
R. "Non-specific" genital infection
S. Swimming-pool conjunctivitis
T. Trachoma
U. Trichomoniasis.

2.50 What is the usual incubation period of syphilis?

V. 24 hours
W. 4 days
X. 1 week
Y. 4 weeks
Z. 3 months.

Answers overleaf

2.47 G. Formation of an inguinal bubo is not characteristic of gonorrhoea whereas it is the first lesion seen in lymphogranuloma venereum, occurs in the majority of cases of chancroid and is commonly present in association with the genital chancre of syphilis. Buboes also typically occur in bubonic plague (*Yersinia pestis*) and in the plague-like disease tularaemia (*Francisella tularensis*).
The other complications used to be not uncommon in gonorrhoea before the antibiotic era but have been markedly decreased in incidence by use of penicillin. Gonococcal endocarditis occurred only rarely before the introduction of antibiotics, and now is never seen.

2.48 O. Chronic infection in the female may spread to involve the Fallopian tubes with subsequent fibrosis and obliteration of the lumen. This may result in tubal pregnancy or sterility. In the male, chronic infection may persist in the posterior urethra and spread to involve the prostate and epididymis. Progressive fibrosis may result in urethral stricture with urinary retention, relieved only by regular dilatation by passage of bougies.

2.49 U. With the exception of trichomoniasis, all the conditions result from infection with TRIC-agents, obligate intracellular parasites of group A chlamydiae. Trichomoniasis is a genito-urinary infection mainly affecting women, caused by a motile pear-shaped protozoan, *Trichomonas vaginalis*. Infection in the female usually affects the vulva, vagina, cervix and urethra, with the production of local inflammation, irritation and a characteristic frothy, yellowish discharge. Male infection causing urethritis, may result from venereal transmission.

2.50 Y. The primary chancre first develops as a small papule at the portal of entry, usually the genitalia or less often the finger or lip, about 3—4 weeks after exposure (10 days—10 weeks).

2.51 Syphilis is a generalized systemic infection. At what stage does *Treponema pallidum* first spread throughout the body?

A. After development of the primary chancre
B. During the latent period
C. Early in the tertiary stage
D. Within hours from the time of exposure
E. When the secondary stage is established.

2.52 Which species of organism infecting a pregnant woman is MOST likely to infect the fetus *in utero?*

F. *Brucella abortus*
G. *Clostridium tetani*
H. *Corynebacterium diphtheriae*
J. *Shigella flexneri*
K. *Treponema pallidum.*

2.53 Which one does *NOT* occur in congenital syphilis?

L. Congenital deformities
M. Hard chancre
N. Intersitial keratitis
O. Positive syphilis serology
P. Still-birth or abortion.

Answers overleaf

2.51 D. *Treponema pallidum* can be detected in the circulation within minutes of inoculation and becomes established throughout the vascular system within hours at most.

2.52 K. Untreated syphilis in pregnancy commonly results in still-birth or, if the infant survives, the stigmata of congenital syphilis. As clinical evidence of early syphilis is often masked in pregnancy, routine serological tests (VDRL screening test, WR and Reiter protein CFT) should be done on all patients in early pregnancy and may be repeated in the third trimester.
Of the other organisms listed only *Br. abortus* is significantly invasive. Unlike in cattle, it does not commonly infect the placenta or fetus and abortion, still birth or fetal infection are rare.

2.53 M. The primary chancre does not develop as the infection is blood-borne via the placenta and the fetal immune system immature and unable to mount a cell-mediated response unless infection occurs very late in pregnancy. If infection occurs during birth, by contact with a very recent primary lesion in the mother, a primary chancre may develop.
The other answers are correct. "Saddle nose", "Hutchinson's teeth" and "sabre-tibia" are not uncommon. Interstitial keratitis is the most common lesion of late congenital syphilis. All serological tests (WR, VDRL, RPCFT, FTA — absorbed and TPI) are usually strongly positive in the blood and CSF.

2.54 Which is a non-venereal form of syphilis?

- Q. Bejel
- R. Pinta
- S. "Trench mouth"
- T. Weil's disease
- U. Yaws.

2.55 Which is the LEAST probable consequence of undiagnosed severe leptospiral infection?

- V. Conjunctival inflammation
- W. Cross infection
- X. Fever
- Y. Jaundice
- Z. Lymphocytic meningitis.

2.56 Which of the following possible explanations accounts for the mechanism of relapse in *Borrelia recurrentis* relapsing fever?

- A. Destruction of leucocytes by leucocidin
- B. Emergence of antigenically distinct mutants
- C. Immunological tolerance
- D. Reinfection from persisting vector
- E. Reticulo-endothelial system blockade.

Answers overleaf

2.54 Q. Bejel is endemic among the Bedouins, mainly contracted in childhood from common drinking vessels. It affects the upper respiratory tract and skin and although gummata are common there is rarely involvement of the cardiovascular and nervous systems. Congenital transmission does not occur. Serological tests for syphilis are positive but the organism is not identical with *T. pallidum.*
The other diseases are all non-venereal: pinta, *Treponema carateum*; "trench mouth", *Borrelia vincenti* plus fusiform bacilli; Weil's disease, *Leptospira icterohaemorrhagiae*; yaws, *Treponema pertenue.*

2.55 W. Communicability between humans is negligible as the organism is excreted only in urine, in which it survives poorly. The others are typically present in severe infections.

Severe infections are usually caused by *Lept. icterohaemorrhagiae* from infected rat urine. The human infection is Weil's disease or haemorrhagic jaundice.

2.56 B. The spirochaete is present in the blood during the fever but can only be found occasionally or in small numbers during the apyrexial period. About the crisis spirochaetes disappear from the blood due to the production of protective antibody specific for the infecting strain. During the apyrexial period some survive within cells in the spleen where they undergo antigenic variation. Subsequent outgrowth of this mutant strain, which is not susceptible to the antibody previously elaborated, is the mechanism of relapse. The final cure after several relapses results from broad immunity afforded by repeated specific responses to multiple mutant strains. This is true for both the European louse-borne and West African tick-borne disease, the latter caused by *B. duttoni.*

2.57 Which organism may produce a latent infection in man lasting for many years?

F. *Bacillus anthracis*
G. *Bordetella pertussis*
H. *Cornebacterium diphtheriae*
J. *Rickettsia prowazeki*
K. *Yersinia pestis.*

2.58 Which is *NOT* a leptospiral disease?

L. Canicola fever
M. Oroya fever (Carrion's disease)
N. Pretibial fever (Fort Bragg disease)
O. "Swine herd's disease"
P. Weil's disease.

2.59 Which of the following is LEAST likely to present clinically as a PUO (Pyrexia of unknown origin)?

Q. Brucellosis
R. Enteric fever
S. Leptospirosis
T. Scarlet fever
U. Subacute bacterial endocarditis.

2.60 Four of these infections commonly present as a pyrexia of unknown origin (PUO): which one does NOT?

V. Chronic meningococcaemia
W. Dysentery
X. Typhoid fever
Y. Primary atypical pneumonia
Z. Q-fever.

Answers overleaf

2.57 J. Individuals recovered from classical typhus fever may harbour *R. prowazeki* for years and then develop a recrudescent form called Brill-Zinsser disease. Symptoms are atypical and there may be no lice infestation to account for the source of infection, as it is of endogenous origin. Chronic carriers allow the continued survival of *R. prowazeki* and may be the source of an epidemic of classical louse-borne typhus fever.

2.58 M. Oroya fever is a severe febrile anaemia which occurs in South America. Patients who survive may later develop a benign skin eruption, Verruga Peruana. Both diseases result from infection by rickettsial-like *Bartonella bacilliformis* transmitted by the sandfly, *Phlebotomus*. Previously fatal, the disease is readily treated with antibiotics and its incidence much reduced by insecticide control of the vector.

2.59 T. Scarlet fever has a sudden onset characteristic of a toxic infection, with the characteristic erythematous rash developing from the second day to suggest the diagnosis. The other conditions may show prolonged fever without localizing signs, thus placing them truly in the category of PUO.

2.60 W. Dysentery is an acute infection caused by shigellae. Mild infections are common but do not produce significant fever. Severe cases (e.g. *Sh. dysenteriae*) are from the outset clearly an intestinal infection and would not present as a PUO.

N. meningitidis commonly invades the blood and occasionally causes persistent bacteraemia in the absence of meningeal localization. *S. typhi* may cause a prolonged pyrexia without intestinal symptoms. *Myco. pneumoniae* and *Cox. burneti* not infrequently produce a PUO. Full investigation of a PUO is a lengthy task and the more common conditions, and those suggested by clinical examination, should be excluded before embarking on the rarities.

2.61 Which one of the following does <u>*NOT*</u> predispose to infection with *Candida albicans?*

A. Diabetes mellitus
B. Immunosuppressive therapy
C. Leukaemia
D. Prolonged treatment with tetracyclines
E. Treatment with any polyene antibiotic.

2.62 Which protozoan infection is mainly associated with eye disease and less commonly with lymphadenopathy and PUO?

F. *Entamoeba histolytica*
G. *Giardia intestinalis*
H. *Plasmodium vivax*
J. *Toxoplasma gondi*
K. *Trichomonas vaginalis.*

Answers overleaf

2.61 E. With the exception of polyene antibiotics, whose main use is the treatment of fungal infections including candidiasis, all the others do predispose to chronic fungal infection due to diminished powers of resistance, or to loss of the protective normal commensal flora. Well known examples of polyene antibiotics are nystatin and amphotericin B. Another anti-fungal antibiotic, 5-fluorocytosine, is not of the polyene group but an anti-metabolite of the pyrimidine base cytosine.

2.62 J. About 700 cases occur annually in England and Wales and of these over 50 per cent affect the eyes and about 30 per cent have fever and lymph gland enlargement. Diagnosis is usually made by serology using a dye test and a complement fixation test. *T. gondi* may be isolated by inoculation of white mice.

3. Virus Infections

3.1 Which statement about yellow fever is FALSE?

A. Diagnosis is based upon intracerebral inoculation of mice
B. Immunity is conferred by a live attenuated vaccine
C. Mosquitos transmit jungle yellow fever but not urban (epidemic) yellow fever
D. Serological diagnosis can be done using neutralization tests in mice
E. The virion has an ether sensitive envelope and an RNA genome.

3.2 In which disease do symptoms appear generally in the first week after exposure to infection?

F. Chickenpox
G. Infectious hepatitis
H. Influenza
J. Mumps
K. Rubella.

3.3 Arrange in sequence the order of spread of poliovirus in an attack of classical paralytic disease, from the time of ingestion of the virus: indicate which phase comes FOURTH.

L. Alimentary phase
M. Ingestion of virus
N. Lymphatic phase
O. Neurological phase
P. Viraemic phase.

Answers overleaf

3.1 C. Both epidemiological types of the disease are caused by the yellow fever virus, the essential difference being that jungle yellow fever involves monkey to man transmission by tree-dwelling mosquitos whereas urban yellow fever results from person to person transmission by the domestic *Aëdes aegypti* mosquito. Anti-mosquito measures in and around cities have eliminated this vector with virtual eradication of the disease. Jungle yellow fever remains a problem in remote areas of Africa and South America where mosquitos perpetuate the disease among the monkey population. Recovery of virus from the blood is possible in the first 5 days of the infection by intracerebral inoculation of mice which subsequently develop an encephalitis. Identification is confirmed by neutralization tests with an emulsion of the mouse brain and specific antisera by intracerebral inoculation of a further mouse. Egg-adapted 17D virus confers a high level immunity for up to 10 years.

3.2 H. The other four diseases have relatively long incubation periods, with symptoms usually appearing in the third week after exposure, between 14—21 days.

3.3 P. After ingestion (M) the virus first multiplies in the pharyngeal and intestinal mucosa (L) before invading the gut-associated lymphoid tissues of the tonsil, Peyer's patches and the regional cervical and mesenteric nodes (N). Thereafter, it is carried in the lymph to the general circulation (P). It breaches the blood-brain barrier, either directly through small vessel walls or via nerves, to reach the vital target cells in the spinal cord (anterior horn cells), the medulla and brain stem, and the motor cortex (O).

3.4 Which one is *NOT* a clinical manifestation of infection with coxsackieviruses?

Q. Aseptic meningitis
R. Enteritis
S. Epidemic myalgia
T. Herpangina
U. Myocarditis and pericarditis.

3.5 Which clinical manifestation of coxsackievirus infection is caused only by certain types of group B.

V. Aseptic meningitis
W. Epidemic myalgia (pleurodynia)
X. Herpangina
Y. Paralytic disease
Z. Respiratory illness.

3.6 Which one of the following RNA viruses is the causal agent of hand-foot-and-mouth disease in man?

A. Coxsackievirus group A
B. Coxsackievirus group B
C. Echovirus
D. Foot-and-mouth disease virus
E. Human rhinovirus.

Answers overleaf

3.4 R. Coxsackieviruses are members of the enterovirus subgroup of picornaviruses but are not associated with enteritis as are the echoviruses. They produce a variety of clinical syndromes which are usually group and even type associated.

Aseptic meningitis (all group B, A7 and A9) may be associated with meningoencephalitis, and a paralytic or nonparalytic poliomyelitis-like disease. *Epidemic myalgia*, known also as pleurodynia or Bornholm disease (group B) is a painful condition affecting the lower, intercostal muscles which resolves within 2 weeks. *Herpangina* (group A) is a feverish inflammation of the throat which affects young children in summer. *Mycardiopathy* (mainly group B, especially B5) may be a benign myocarditis or pericarditis in the adult or a fatal congenital or perinatal myocarditis which occurs in small but very serious outbreaks, usually transmitted by an infected mother. Throat swab and faeces are suitable specimens for virus isolation.

3.5 W. Also known as Bornholm disease, it is caused by group B viruses alone. The condition occurs in epidemics characterized by fever, severe pain in the lower intercostal muscles, particularly when coughing, and occasionally complicated by mild meningitis, orchitis and rarely diabetes due to involvement of the pancreas. Usually however, the disease is mild and self limiting, lasting up to 2 weeks.

3.6 A. Coxsackieviruses A16, or less commonly A4, A5, A9 and A10, are associated with a disease affecting young children, characterized by fever plus a painful vesicular rash on the hands, feet, oral mucosa and buttocks. The virus is present in the vesicular lesions and can also be recovered from faeces and throat washings. The disease must not be confused with foot-and-mouth disease, caused by animal-type rhinovirus, which affects cattle, sheep, pigs and goats and only rarely is transmitted to man from infected animals: human disease is characterized by vesicular lesions on the hands, feet and mouth.

3.7 **What is the usual incubation period of hepatitis B (serum hepatitis)?**

F. 24 hours
G. 4 days
H. 1–2 weeks
J. 1–2 months
K. 2–4 months.

3.8 **Which one does *NOT* occur in the course of an infection with measles virus?**

L. Development of nuclear and cytoplasmic inclusion bodies
M. Induction of chromosomal abnormalities in blood leukocytes
N. Persistence of virus with recurrence of acute disease
O. Production of interferon and protective specific antibody
P. Stimulation of delayed hypersensitivity to the virus.

3.9 **Which is *NOT* a property of respiratory syncytial virus?**

Q. It is a member of the paramyxoviruses
R. It is identified by the haemagglutination-inhibition test
S. It produces giant cells and syncytia in HeLa cell culture
T. The helical capsid contains single-stranded RNA
U. The virion is ether sensitive and extremely labile.

Answers overleaf

3.7 K. Serum hepatitis has a much longer incubation period than does infectious hepatitis (2—6 weeks). The virus can be detected in the blood during the incubation period and the acute phase of the disease.

3.8 N. Recurrences of measles do not occur as the development of acquired active immunity gives permanent protection from reinfection and the measles virus does not persist in the body in a latent form capable of causing recrudescence of acute disease. Rarely, the virus may persist in an altered, possibly incomplete form when it causes the slowly progressive subacute sclerosing panencephalitis. Large multinucleate giant cells with intranuclear and cytoplasmic inclusion bodies are produced both in tissue culture and in natural infection. They may be demonstrated in nasal secretions, sputum and in gut-associated lymphoid tissues. Chromosomal abnormalities may be seen in peripheral blood leucocytes in the course of infection as well as in cultured cells. Interferon is produced by infected cells. Specific antibody produced in response to infection is active in prevention of reinfection and may be detected by neutralization and complement fixation tests. Although measles is notorious for diminishing the cell mediated immune reaction (e.g. loss of tuberculin type hypersensitivity and reduction in lymphocyte transformation) it does induce a state of delayed hypersensitivity which may be demonstrated by a skin test. Furthermore, the development of measles giant-cell pneumonia in persons with cell-mediated deficiency underlines its protective role.

3.9 R. Respiratory syncytial virus lacks the haemagglutinin characteristic of the myxo- and paramyxoviruses, so that the HI-test is inappropriate and instead, neutralization and complement fixation tests are used. The virus is very widely distributed, infecting 80 per cent of children by the age of 4 years. It is the commonest virus reported apart from influenza type A and is the single most important cause of potentially fatal bronchiolitis and pneumonia which occurs in babies during the first year of life.

3.10 Which one is *NOT* associated with the "rubella syndrome"?

V. Cataract
W. Deafness
X. Giant-cell pneumonia
Y. Microcephaly
Z. Patent ductus arteriosus.

3.11 Which slow virus disease is said to have been transmitted by cannibalism?

A. Creutzfeldt-Jacob disease
B. Kuru
C. Progressive multifocal leucoencephalopathy
D. Scrapie
E. Subacute sclerosing panencephalitis.

3.12 Which of the following viruses usually causes a generalized infection in man?

F. Cowpox
G. Molluscum contagiosum
H. Orf
J. Smallpox
K. Vaccinia.

Answers overleaf

3.10 X. Giant-cell pneumonia occurs as a rare complication of measles in debilitated or immune-deficient children. Women who contract rubella in the first eight weeks of pregnancy have 90 per cent chance of fetal infection and 50 per cent chance that the baby will be born with developmental abnormalities. Active infection may persist in the infant after birth despite natural active immunity.

3.11 B. *Kuru* is a chronic, slowly progressive, fatal spongiform encephalopathy caused by a small, poorly characterized transmissible agent. The disease is limited to natives of New Guinea who, until prohibitive legislation was introduced in 1957, had practised cannibalism of dead relatives. Symptoms of progressive cerebellar ataxia appear after an incubation period of from months to over 40 years. The outlawing of cannibalism has reduced the incidence of the disease.

Creutzfeldt-Jacob disease is a similar disease of a much wider geographical distribution unrelated to cannibalism. *Scrapie* is another similar spongiform encephalopathy which affects sheep. *Progressive multifocal leucoencephalopathy* is a human demyelinating disorder. Papovaviruses (e.g. SV40-PML virus and JC virus) have been isolated from cases of this disease. *Subacute sclerosing panencephalitis* is a neurological disease which occurs as a very rare, extremely delayed complication of measles.

3.12 J. All these viruses belong to the poxvirus group. With the exception of smallpox, infection is limited to the skin.

3.13 Which one of the following viruses is <u>*NOT*</u> a member of the poxvirus group?

L. Chickenpox virus
M. Cowpox virus
N. Molluscum contagiosum virus
O. Smallpox virus
P. Vaccinia virus.

3.14 Which one of the following statements about variola minor (alastrim) is FALSE?

Q. Alastrim strains never cause variola major
R. Form of mild smallpox in the partially immune
S. Incubation period of usually 12 days
T. Man is the natural host and sole reservoir
U. The virus can be isolated on chorio-allantoic membrane of the chick embryo.

3.15 Against which virus infection has the prophylactic administration of methisazone, to close contacts of an active case, been successful in averting or modifying an attack of the disease?

V. Hepatitis A
W. Hepatitis B
X. Lassa fever
Y. Smallpox
Z. Yellow fever.

Answers overleaf

3.13 L. Chickenpox (varicella) virus is one of the herpes viruses and not a poxvirus. Chickenpox is a usually mild but highly infectious disease of childhood which produces repeated crops of characteristic skin spots, mainly over the trunk and face. The virus is transmitted by droplet spread from the upper respiratory tract and by contamination from discharging lesions. Second attacks are rare due to acquired immunity but latent infection persisting after a clinical attack of chickenpox may be reactivated in the adult, at a time of waning immunity, to produce herpes zoster (shingles) which is a further but less active source of chickenpox. The virions are 100—150 nm icosahedrons showing cubic symmetry and are easily distinguishable from smallpox virions which are 200—300 nm brick shaped particles, with complex symmetry. Thus electron microscopy of fluid from a lesion offers the most rapid differential diagnosis where the probability of smallpox is considered.

3.14 R. The mild character of alastrim is the direct result of the low level of virulence for man of the variola minor strains and is not simply the expression of classical smallpox strains in persons with a degree of immunity due to prior immunization with vaccinia. Secondary cases always develop variola minor. Variations in the severity of classical smallpox is certainly related to the degree of protection afforded by vaccination but secondary cases arising from mild classical smallpox develop variola major.

3.15 Y. Methisazone (N-methylisatin-β-thiosemicarbazone) is known to reduce the incidence of smallpox and alastrim if administered within one or two days of contact. It is not effective in established smallpox but has proved of value in the treatment of *vaccinia gangrenosa* and *eczema vaccinatum*.

3.16 Which one is *NOT* a clinical manifestation of infection with adenovirus?

A. Acute respiratory disease
B. Epidemic keratoconjunctivitis
C. Intussusception of infancy
D. Pharyngoconjunctival fever
E. Subacute sclerosing panencephalitis.

3.17 Which one is *NOT* caused by herpes simplex virus?

F. Acute gingivostomatitis
G. Cold sores
H. Kaposi's varicelliform eruption
J. Meningoencephalitis
K. Shingles.

3.18 Which clinical condition do you associate with the herpes simplex virus?

L. Croup
M. Herpangina
N. Herpes zoster
O. Keratitis
P. Pericarditis.

Answers overleaf

3.16 E. This progressive cerebral degeneration occurs rarely in children and young adults following an attack of measles. The mechanism is unknown but the virus is thought to persist in the CNS.
Acute respiratory disease (types 4 and 7) includes inflammations affecting the pharynx, larynx, trachea, bronchi and lungs. *Epidemic kerato-conjunctivitis* (type 8) has been associated with employment which cause corneal abrasions, hence "shipyard eye". *Intussusception* (types 1, 2, 3 and 5) is probably the result of hyperplasia of the gut-associated lymphoid tissue due to infection. The enlarged follicles protrude into the lumen to initiate the intussusception. *Pharyngo-conjunctival fever* may be produced by several types, conjunctivitis being prominent in type 3 infections.

3.17 K. Shingles is caused by reactivation of varicella-zoster virus lying dormant for many years following childhood chickenpox — hence the name, varicella-zoster virus. Although the disease is not transmissible, it may be the source of an outbreak of chickenpox.

3.18 O. Herpetic keratitis results from primary infection of the cornea, possibly secondary to a corneal abrasion. Infection may produce either a dendritic or amoeboid ulcer which are susceptible to idoxuridine. The virus may persist as a latent infection to recur at some later date. Repeated recurrences cause progressive scarring with impairment of vision or even blindness. Parainfluenza virus is the commonest cause of croup in infants and young children although the syndrome has many aetiologies. Herpangina is caused by coxsackieviruses of group A. Herpes zoster or shingles is due to reactivation of the varicella-zoster virus. Pericarditis and myocarditis have been associated with coxsackieviruses of group B.

3.19 Which clinical syndrome is *NOT* a manifestation of infection with a virus of the herpes group?

Q. Burkitt's lymphoma
R. Chickenpox
S. Cytomegalic inclusion disease
T. Marek's disease
U. Molluscum contagiosum.

3.20 Infection in early pregnancy with which virus may result in serious congenital defects or death of the fetus?

V. Cytomegalovirus
W. EB virus
X. Group A coxsackievirus
Y. Parainfluenza virus
Z. Rhinovirus.

3.21 Which virus is responsible for an infectious mononucleosis-like syndrome but *without* heterophile antibodies?

A. Cytomegalovirus
B. EB virus
C. Group A coxsackievirus
D. Parainfluenza virus
E. Rhinovirus.

Answers overleaf

3.19 U. Molluscum contagiosum virus belongs to the poxvirus group which includes the viruses of smallpox, vaccinia, cowpox, orf and "milker's nodules" (paravaccinia).

3.20 V. The widely known relationship of congenital rubella with fetal abnormalities has tended to overshadow the associations of congenital cytomegalic inclusion disease with developmental defects. Latent infection in the mother may be transmitted to the fetus which may develop hepatitis with hepatosplenomegaly, thrombocytopenia and jaundice as well as very serious central nervous system defects which include microcephaly, mental and motor retardation, optic atrophy and chorioretinitis. Diagnosis of infection is established by either isolation of the virus from urine, saliva or salivary gland biopsy or by detection of a rising titre of specific antibodies by complement fixation or neutralization tests but the clinical significance is always open to question due to the high prevalence of infection in healthy individuals.

3.21 A. Cytomegalovirus infection may produce a high spiking fever which persists for several weeks and is accompanied by a mononucleosis of 2000—20,000 per mm^3 with atypical lymphocytes suggestive of glandular fever but with a negative Paul-Bunnell reaction. The syndrome has been particularly associated with fresh blood transfusion some 30—40 days previously. It is thought the virus is derived from a latent infection in the donor, remaining infectious so long as the blood is not refrigerated which would inactivate it. EB herpes virus is associated with infectious mononucleosis with 50—90 per cent of patients developing heterophile agglutinins for sheep erythrocytes.

3.22 Which virus has been implicated as the causal agent of infectious mononucleosis (glandular fever)?

F. Adenovirus
G. Epstein-Barr virus
H. Hepatitis A virus
J. Herpes simplex virus
K. Varicella-zoster virus.

3.23 Which statement about rabies is FALSE?

L. Prophylactic immunization is available to those at risk
M. The usual incubation period is 1—2 months
N. The virus can infect a wide range of host animals
O. The virus is excreted in saliva and is transmitted by biting
P. Virus growth is restricted to the brain.

Answers overleaf

3.22 G. Infectious mononucleosis occurs in three forms: anginose (sore throat), glandular (enlarged lymph nodes and spleen) and febrile (fever). There is usually a marked monocytosis with atypical lymphocytes in the blood (must be differentiated from leukaemia). From 50—80 per cent produce diagnostic heterophile antibodies (Paul-Bunnell test) and may form antibodies to EB virus. It mainly affects the 14—25 age group and is possibly transmitted by kissing. EB virus may also be associated with Burkitt's lymphoma and post-nasal carcinoma.

3.23 P. Virus in fact is widely disseminated throughout the body, especially in nervous tissue, but can multiply in other tissues (e.g. salivary gland, pancreas, muscle and connective tissue).
The fox is the main natural reservoir in Europe, rabid foxes transmitting the disease to wild, farm and domestic animals, especially dogs and cats. In South America the vampire bat is an important exotic host as it may develop latent infection.
Prophylactic immunization may be offered to persons at high risk (e.g. person bitten in a rabies area and port veterinary officers).

3.24 Which statment about Lassa fever is FALSE?

Q. Close human contacts have a high risk of infection
R. Lassa virus is a member of the arenavirus group
S. Mild or subclinical infections do not occur
T. The natural reservoir is a wild rat indigenous to Africa
U. Transmission usually occurs through contamination of food and dust by infected rat urine.

Answers overleaf

3.24 S. Although the severe form of Lassa fever has been well publicized, the vast majority of cases in the local population are mild or inapparent, recognized only by positive serology during surveys in endemic areas. The disease was first recognized in West Africa in 1969 and has since occurred in epidemics and as hospital outbreaks. The African wild rat *Mastomys natalensis* is the only known reservoir but the risk of transfer to a susceptible European rodent should not be excluded. The disease develops, after 7—10 days incubation period, as an acute fever with general malaise, muscle pains, vomiting and diarrhoea and profound prostration. Conjunctival injection, pharyngeal inflammation and ulceration and generalized lymphadenopathy commonly occur. Fulminating cases develop haemorrhagic lesions within 5 days and 50 per cent rapidly die in shock. Diagnosis is based on demonstration of the virus in blood, urine and throat washings by a rapid indirect immunofluorescence test or by isolation of the virus in cell culture, but this is slower, taking 5 days. Serological diagnosis with paired sera is less satisfactory as significant titres rarely occur until after 15 days. Cross reactions with other arena viruses reduces reliability, and it is mainly of value in epidemiological surveys in endemic regions.

4. Laboratory Diagnosis of Infection

4.1 Which test does NOT depend upon diffusion in gel?

A. Elek plate test for *C. diphtheriae* toxigenicity
B. Measurement of MLD or LD_{50}
C. Precipitin reaction by Mancini method
D. Precipitin reaction by Ouchterlony technique
E. Test for antimicrobial sensitivity by multodisk.

4.2 Which factor has LEAST effect in preventing recovery of bacteria from blood cultures taken from a septicaemic patient?

F. Culture methods employed
G. Fastidiousness of the organism
H. Presence of antibodies, complement and antimicrobials
J. Presence of a high leucocytosis
K. Volume of blood cultured.

4.3 Which investigation establishes the diagnosis of subacute bacterial endocarditis?

L. ASO titre
M. Detection of antibody to *Strep. viridans*
N. Serial blood culture
O. Throat swab
P. Tooth socket swab.

4.4 Which laboratory diagnosis can be made within 5 minutes?

Q. Diphtheria
R. Enteric fever
S. Septicaemia
T. Vincent's angina
U. Whooping cough.

Answers overleaf

4.1 B. Bacterial virulence may be measured by determining the lethal effect of cells or toxin, injected into animals. Dose-response curves are constructed for the entire group (minimum lethal dose — MLD) or for 50 per cent of the group (LD_{50}). Thus the extremes of resistance are excluded, producing a more uniform test model. The other tests are based on diffusion in gel.
A, C and D are gel-precipitin reactions and E diffusion of antibiotic from an impregnated disc.

4.2 J. Leucocytosis does not affect recovery of the organism on culture. Adequate culture methods are essential and the normal range may be supplemented by special enriched media for fastidious organisms and the normal 10—14 days incubation period lengthened. Inhibitory substances are inactivated by dilution and by addition of specific antagonists (penicillinase, PAB, etc.).

4.3 N. S.B.E. is characterized by arterial emboli with *intermittent* bacteraemia. *Serial* blood culture generally provides the bacterial diagnosis, which forms the basis for rational antibiotic therapy. With the exception of specimens taken during cardiac surgery or at autopsy, blood culture is the only direct method of diagnosis. There is an increasing incidence of organisms other than *Strep. viridans* causing this condition (e.g. enterococci).

4.4 T. Demonstration of fusiform bacilli and spirochaetes in large numbers in a Gram-film confirms the clinical diagnosis. The other infections require cultural identification of the causal organism, and even the rapid fluorescent-labelled antibody method for diagnosing whooping cough would take about 30 minutes—1 hour.

4.5 Which method permits the MOST rapid diagnosis of whooping cough?

V. Cough plate culture
W. Detection of specific agglutinins
X. Differential blood count
Y. Fluorescent antibody technique
Z. Per-nasal pharyngeal swab culture.

4.6 In which infection is isolation of the causal organisms greatly facilitated by the use of per-nasal swabs, the cough plate technique and selective culture media?

A. Meningococcal rhinopharyngitis
B. Miliary tuberculosis
C. Pertussis
D. Staphylococcal bronchopneumonia
E. Vincent's infection.

4.7 Isolation of which respiratory pathogen is facilitated by the use of a Bordet-Gengou plate?

F. *Bordetella pertussis*
G. *Coxiella burneti*
H. *Haemophilus influenzae*
J. *Mycoplasma pneumoniae*
K. *Streptococcus pneumoniae.*

4.8 Which result does *NOT* support a provisional diagnosis of primary atypical pneumonia due to mycoplasma infection?

L. Detection of causative agent by immunofluorescence
M. Detection of specific complement fixing antibody
N. Isolation of *Mycoplasma hominis* from sputum
O. Rising titre of agglutinations against streptococcus MG
P. Rising titre of cold agglutinins against group O erythrocytes.

Answers overleaf

4.5 Y. Nasopharyngeal swab smears stained with specific *Bordetella pertussis* fluorescein-labelled antibody give an immediate identification of the causal organism. Standard isolation procedures take at least 2 days, specific agglutinins appear only after the third week of the illness and their detection is then of little practical value. The differential WBC count can never do more than provide confirmatory evidence by showing the characteristic leucocytosis of 15,000—30,000, with a relative and absolute lymphocytosis, towards the end of the catarrhal stage.

4.6 C. *Bordetella pertussis* is a fastidious pathogen which requires special culture media and is readily overgrown by contaminating commensals of the nasopharynx. Earlier methods for its isolation depended on the use of Bordet-Gengou cough plate but this has now been largely replaced by per-nasal swabs, cultured on Bordet-Gengou or alternative special media, incorporating penicillin to inhibit any chance contaminants. Successful isolation is most likely during the early catarrhal phase of the disease, before secondary opportunists become prevalent.

4.7 F. Bordet-Gengou agar contains potato, blood and glycerol which promote the growth of *B. pertussis*. Posterior pharyngeal swabs taken early in the disease give good isolation. Addition of penicillin to the medium improves the isolation rate by inhibiting commensals thus selecting the pathogen.

4.8 N. The causal agent of primary atypical pneumonia is *M. pneumoniae* (Eaton's agent) and not *M. hominis* which is found in the genital tract and is of very doubtful pathogenicity. The other diagnostic procedures provide supportive evidence of infection and the specific immunofluorescence and CF tests are of diagnostic value.

4.9 Which positive test provides a rapid and fairly reliable means for distinguishing infections from mere colonization with *Candida albicans?*

Q. Demonstration of candida precipitins by counter-immuno-electrophoresis
R. Detection of specific cell-mediated immunity *in vitro*
S. Microscopy of Gram-stained film of sputum
T. Sabouraud's plate culture of sputum
U. UV microscopy using direct fluorescent-antibody technique.

4.10 Which one of the following is always of clinical significance when present in sputum?

V. *Candida albicans*
W. *Haemophilus influenzae*
X. *Klebsiella pneumoniae*
Y. *Mycobacterium tuberculosis*
Z. *Staphylococcus aureus.*

4.11 What is the minimum number of tubercle bacilli per ml of sputum necessary for their detection as acid-fast bacilli in a routine Ziehl-Neelsen film?

A. 10
B. 10^2
C. 10^3
D. 10^5
E. 10^{10}.

Answers overleaf

4.9 Q. Detection of a specific humoral or cell-mediated immune response has been interpreted as evidence of *Candida albicans* infection although only 50 per cent of proven septicaemias in immunosuppressed subjects give positive reactions. Tube precipitin tests and specific transformation of lymphocytes take about 2 days whereas counter-immuno-electrophoresis gives a positive result within 2 hours which is important in such a serious infection.
The other tests listed cannot differentiate infection from carriage although the presence of mycelial growth is more indicative of infection. Dubiety still exists over the diagnostic value of candida precipitin reactions.

4.10 Y. Detection of "acid fast" bacilli in sputum is sufficient to warrant making a presumptive diagnosis of pulmonary tuberculosis. Culture of *M. tuberculosis* is taken as firm evidence of pulmonary tuberculosis as there is no carrier state with this organism.
The other organisms all from time to time form part of the commensal flora of the upper respiratory tract, particularly in the mouth, where they may successfully colonize the mucous membranes without causing harm. Overgrowth of the normal mixed commensal flora with one or several of these organisms may be encouraged by poor oral hygiene and the use of broad spectrum antibiotics, and their presence in sputum may result simply from contamination of the specimen with saliva.

4.11 D. It has been estimated that numbers below 10^5 are highly unlikely to be detected in the average clinical laboratory and auramine stained films examined by ultraviolet microscopy show no significant improvement in sensitivity.

4.12 What does a positive tuberculin test indicate in a seemingly healthy middle-aged man who has never been vaccinated with BCG?

F. He does not require BCG vaccination
G. He does not suffer from active tuberculosis
H. He is completely immune to tuberculosis
J. He requires BCG vaccination
K. He suffers from active tuberculosis.

4.13 Which one of the following is diagnostic of genito-urinary tuberculosis?

L. Abacterial pyuria
M. Acid-fast bacilli in urinary sediment
N. Isolation of *M. tuberculosis* from *one* urine sample
O. Negative tuberculin reaction
P. Positive tuberculin reaction.

4.14 Which medium is specifically designed to aid the isolation of *Staphylococcus aureus* from heavily contaminated specimens, such as faeces?

Q. Blood agar
R. Blood — tellurite agar
S. Crystal violet — blood agar
T. Mannitol — salt agar
U. MacConkey's bile salt agar.

Answers overleaf

4.12 F. A positive reaction to a tuberculin test in a person who has not received BCG indicates the presence of a tuberculous infection, but does not indicate whether the lesion is active or inactive. It does confirm that the individual is hypersensitive to tuberculin and therefore does not require BCG vaccination.
Tuberculin hypersensitivity confers a degree of resistance to tuberculosis but does not provide solid immunity. Active tuberculosis could subsequently result from reactivation of a dormant primary focus or from reinfection from inhalation of a large number of virulent organisms.

4.13 N. This alone is absolute proof of tuberculous infection. Repeated demonstration of acid-fast bacilli at best provides a provisional diagnosis, to be confirmed by positive culture, as *M. smegmatis* may be present. A positive tuberculin reaction does not differentiate the hypersensitivity induced by childhood primary tuberculosis from a current infection as even the degree of response is very variable: nor does a negative reaction exclude the disease, as a recent infection might not yet have converted and overwhelming infections may show anergy, as would a deficiency of cell-mediated immunity. Tuberculosis must always be excluded in persons with abacterial pyuria although it is more often due to other causes (e.g. undisclosed prior antibiotic therapy for what is truly a coliform infection).

4.14 T. High concentration (10 per cent) of sodium chloride is inhibitory to most species other than staphylococci. Differentiation of *Staph. aureus* from *Staph. albus* is aided by colour change of mannitol-fermenting colonies which indicates possible *Staph. aureus*, and these can subsequently be confirmed by demonstration of coagulase production.

4.15 **What is the value of sodium desoxycholate in the examination of faeces in a case of diarrhoea?**

V. Encourages growth of enteropathogenic *Escherichia coli*
W. For maintenance of salmonella type cultures
X. Medium for faecal coliform count
Y. Medium for isolation of *Vibrio cholerae*
Z. Medium for selective culture of salmonella and shigella.

4.16 **Examination of faeces from a woman complaining of fever, mild diarrhoea and severe abdominal pain resulted in isolation of a motile, non-lactose fermenting coliform which fermented glucose and mannite with formation of acid but no gas, and was oxidase, urea and indole negative. Which intestinal pathogen is consistent with these findings?**

A. *Salmonella typhi*
B. *Salmonella paratyphi B*
C. *Shigella flexneri*
D. *Shigella sonnei*
E. *Vibrio cholerae.*

4.17 **Which diagnostic laboratory test is MOST likely to be positive in the first week of enteric fever?**

F. Serial blood culture
G. Stool culture
H. Urine culture
J. Widal reaction
K. None of these.

Answers overleaf

4.15 Z. Sodium desoxycholate is incorporated in a nutrient agar plate used for primary isolation of salmonella and shigella. It acts by inhibiting the growth of commensal coliforms while the pathogens are unaffected and are easily identified. Prior incubation of faeces in a liquid enrichment broth (e.g. tetrathionate or selenite broth) before subculture to a desoxycholate citrate agar (DCA) plate will increase the proportion of pathogens and facilitate their isolation from specimens where they are few in number.

4.16 A. *Salmonella typhi* is exceptional amongst salmonella in its failure to produce gas on fermentation. Shigellae are excluded by positive motility. *Vibrio species* are indole positive and produce gas on fermentation of glucose and possibly lactose.

4.17 F. The septicaemic phase occurs in the first week and repeated blood cultures usually yield a positive isolation from one or more bottles. The intestine is not significantly involved until reinfection from the septicaemia via the bloodstream or indirectly from infected biliary secretions, giving a positive stool culture in the second week. Urine culture becomes positive in the third week and positive serology is possible from the second week onwards, serial examination giving a diagnostic rising titre. Although this pattern is commonly found it is not constant and all tests should be used from the beginning to ensure the earliest possible diagnosis of this serious infectious disease.

4.18 What procedure is required to establish the identity of a presumptive salmonella isolate?

L. Antigenic analysis
M. Biochemical reactions
N. Chemical analysis
O. Gram reaction
P. Motility test.

4.19 Which gastrointestinal disease is associated with a faecal leucocytosis of which more than 80 per cent are mononuclears?

Q. Infantile gastroenteritis (*Escherichia coli* 0119)
R. Salmonellosis (*S. typhimurium*)
S. Shigellosis (*Sh. flexneri*)
T. Staphylococcal food poisoning (*Staph. aureus*)
U. Enteric fever (*S. typhi*).

4.20 What is the significance of a positive reaction with ONPG (O-nitrophenyl-β-D-galactopyranoside) in the investigation of an enteric organism?

V. Confirms the presence of a salmonella
W. Excludes the possibility of enteropathogenic *E. coli*
X. Indicates the organism ferments lactose
Y. Proves it is not a shigella
Z. Shows the organism is oxidase-positive.

Answers overleaf

4.18 L. Salmonella are identified by their complement of somatic and flagellar antigens according to the Kauffman-White scheme. Positive slide agglutinations with polyvalent salmonella H and O antisera are followed by testing with group specific O-factor antisera (usually A, B, C, D or E) and non-specific and specific phase H antisera. Where the organism is in the non-specific phase 2, the specific phase 1 may be recovered in a Craigie tube. In this apparatus non-specific phase 2 antiserum immobilizes phase 2 forms in an inner tube but allows the motile phase 1 population to migrate to the outer tube where it may be recovered in a relatively pure (genetically speaking) state.

4.19 U. In patients with established typhoid fever, examination of a methylene-blue stained wet film reveals numerous leucocytes under the low power objective. A differential white cell count done under the high dry (x40) objective shows that more than 80 per cent are mononuclear cells. This is in sharp contrast to the enormous predominance of neutrophils in infantile gastroenteritis, shigellosis and salmonella food poisoning. Staphylococcal food poisoning may show no increase in leucocytes unless the condition is a staphylococcal enterocolitis (often after broad spectrum antibiotics) when polymorphs and masses of staphylococci are seen.

4.20 X. It shows the organism produces the enzyme B-galactosidase as revealed by development of a yellow colour due to release of O-nitrophenol from the substrate. The test is used to differentiate late lactose fermentation from true lack of ability to ferment lactose. Salmonella may thus be eliminated but several shigella, especially *Sh. sonnei* are ONPG positive.

4.21 **For selective isolation of which specific intestinal pathogen is thiosulphate citrate, bile salts, sucrose agar (TCBS) recommended?**

A. *Entamoeba histolytica*
B. Enteropathogenic *Escherichia coli*
C. *Salmonella typhi*
D. *Shigella dysenteriae*
E. *Vibrio cholerae.*

4.22–24 **For each culture medium on the left, select the MOST suitable organism from the list on the right.**

4.22 Lowenstein-Jensen medium
4.23 Mannitol-salt agar
4.24 Tellurite-blood agar

F. *Acinetobacter anitratus*
G. *Bordetella pertussis*
H. *Corynebacterium diphtheriae*
J. *Mycobacterium tuberculosis*
K. *Staphylococcus aureus.*

4.25–27 **For each organism on the left, select the MOST useful selective medium from the list on the right.**

4.25 *Corynebacterium diphtheriae*
4.26 *Salmonella typhimurium*
4.27 *Staphylococcus aureus.*

L. Blood-tellurite agar
M. Deoxycholate citrate agar
N. Lowenstein-Jensen medium
O. Mannitol-salt agar
P. Sabouraud's agar.

Answers overleaf

4.21 E. This medium was introduced for selective isolation of vibrios responsible for cholera (*V. cholerae* and its El Tor biotype) and food poisoning (*V. parahaemolyticus*). Vibrios differ from the coliforms in preferring an alkaline environment with a high salt content. Such conditions are provided by TCBS medium which suppresses the growth of most other organisms in faeces and therefore greatly aids primary isolation of vibrios. Colonies of *V. cholerae* and El Tor biotype form butter yellow colonies within 24 hours and *V. parahaemolyticus* can be distinguished by forming colonies with greenish-blue centres.

4.22—24 4.22 J: 4.23 K: 4.24 H.
Lowenstein-Jensen slopes are widely used for primary isolation of mycobacteria. Glycerol incorporated in the medium differentiates the abundant (eugonic) growth of the human strain, *M. tuberculosis*, from the scantier (dysgonic) growth of the bovine strain, *M. bovis*. Mannitol salt agar is used as a selective medium for *Staph. aureus* allowing its easy recovery from heavily mixed or contaminated specimens. *Staph. aureus* is relatively salt resistant whereas most other organisms ae inhibited by high salt concentration and most strains rapidly ferment mannitol giving a colour change (cf. lactose in MacConkey medium), whereas non-pathogenic *Staph. albus* commonly fails to do this. Tellurite-blood agar is a selective medium for corynebacteria, and allows the development of distinctive colonial morphology for gravis, intermedius and mitis strains of *C. diphtheriae.*

4.25—27 4.25 L: 4.26 M: 4.27 O.
Blood-tellurite not only selects *C. diphtheriae* but also helps to differentiate gravis, intermedius and mitis strains by their characteristic colonial morphology and differences in haemolysis. DCA aids recovery of salmonella from faecal specimens by selectively inhibiting commensal coliforms. MSA inhibits faecal coliforms due to the high salt content and mannitol fermentation differentiates *Staph. aureus* from non-fermenting *Staph. albus.*

4.28 Which one of the following is *NOT* increased in a case of acute pyogenic meningitis?

Q. CSF cell count
R. CSF glucose concentration
S. CSF pressure
T. CSF protein concentration
U. Peripheral leucocyte count.

4.29 Which is the MOST probable causal organism in a case of meningitis where the CSF shows the following changes: CSF slightly cloudy but colourless; protein raised, sugar and chloride lowered; cell count 400/c.mm with lymphocytes greatly exceeding neutrophil polymorphs?

V. *Escherichia coli*
W. *Haemophilus influenzae*
X. *Mycobacterium tuberculosis*
Y. *Neisseria meningitidis*
Z. *Streptococcus pneumoniae.*

4.30 Which organism gives an improved isolation rate from clinical specimens which have been held at 4°C for several days?

A. *Bacteroides fragilis*
B. *Listeria monocytogenes*
C. *Neisseria gonorrhoeae*
D. *Neisseria meningitidis*
E. *Streptococcus pyogenes.*

Answers overleaf

4.28 R. In cases of pyogenic meningitis the CSF glucose level is markedly decreased or sugar may even be completely absent, due to the combined glycolytic effect of the bacteria and enzymes released from damaged cells. The cell count is usually between 500 and 20,000 polymorphs per cu mm. The peripheral WBC count may show a fairly high neutrophil leucocytosis.

4.29 X. The two main differential features of CSF profile are the naked eye appearance of only slight cloudiness and the only moderately raised cell count with predominance of lymphocytes. The other forms of meningitis are all pyogenic and would probably have a cell count well in excess of that present, and the differential count would show they were almost exclusively neutrophil polymorphs. Confirmation requires Gram and Ziehl-Neelsen films, and appropriate culture.

4.30 B. For an as yet unknown reason, *Listeria monocytogenes*, present in small numbers in a clinical specimen, may fail to grow on initial culture whereas it can be recovered by subsequent culture of the same material stored in a refrigerator at 4°C for days or even weeks.

4.31 Which laboratory diagnostic test for syphilis is of LEAST value in the tertiary stage of the disease?

F. Darkground microscopy of material from the lesions
G. Histological examination of affected tissues
H. Reiter protein complement-fixation test
J. *Treponema pallidum* immobilization test
K. Wassermann reaction or VDRL slide test.

4.32 Which test, when positive, is the MOST reliable indication of past or present infection by *Treponema pallidum*?

L. FTA-absorbed test
M. Kahn flocculation reaction
N. Lange's colloidal gold test
O. VDRL slide test
P. Wassermann reaction.

4.33 Which serological test is the MOST reliable indicator of latent syphilis many years after contracting the disease?

Q. Kahn flocculation test
R. Reiter protein complement-fixation test
S. *Treponema pallidum* immobilization test
T. Venereal disease research laboratory screening test
U. Wassermann reaction.

4.34 The Wassermann reaction is POSITIVE if antibody coated sheep erythrocytes (added in the second stage of the reaction) show which one of the following?

V. Agglutination
W. Erythrocyte binding of complement
X. Lysis
Y. No agglutination
Z. No lysis.

Answers overleaf

4.31 F. Demonstration of the characteristic spirochaete by dark-ground microscopy is possible only in the primary and secondary stages of the disease when spirochaetes are abundant in the exudates from ulcerated lesions of the skin and mucosae. After the secondary stage subsides, the treponemes become few in number, although they may still be demonstrable histologically in and around affected blood vessels. Serological tests form the basis of laboratory diagnosis of syphilis.

4.32 L. The FTA-absorbed test is based upon specific antibody to *T. pallidum* and once positive may persist for life. False positives found when using neat serum are excluded by 200 times dilution before the test. The other tests listed are all based upon the non-specific heterophile WR-type antibody, or reagin: false biological positives are not uncommon in pregnancy and in the course of other chronic infections, such as leprosy, tuberculosis, etc.

4.33 S. The *Treponema pallidum* immobilization test is based upon testing the ability of a patient's sera to immobilize a motile suspension of *Treponema pallidum* (Nichol's strain) due to the presence of virtually specific antibody. A simpler and almost as reliable procedure is the fluorescent treponemal antibody (FTA) test based also upon detection of the same specific antibody.

4.34 Z. The WR is a complement-fixation test used to detect heterophile antibody produced in response to syphilis. The test has two stages. First, the patient's serum is incubated with the cardiolipin antigen in the presence of a measured amount of guinea pig complement. Seçondly, a haemolytic system of sensitized sheep erythrocytes (erythrocytes plus specific antibody to them) is added. If complement has been bound in the first stage, it is not free to lyse the haemolytic system and the test is *positive*. If no antigen-antibody union occurred in stage one, complement remains free to lyse the erythrocytes in stage two and the test is negative.

4.35 Which skin test is used to aid diagnosis of *Lymphogranuloma venereum?*

A. Casoni reaction
B! Frei reaction
C. Matsuda reaction
D. Schwartzman reaction
E. Weil-Felix reaction.

4.36 All of the following disease may present clinically as a pyrexia of unknown origin (PUO). Which one is diagnosed by a sheep erythrocyte agglutination test?

F. Brucellosis
G. Infectious mononucleosis
H. Syphilis
J. Tuberculosis
K. Typhoid fever.

4.37 Which test is MOST likely to establish a diagnosis of leptospiral infection in the first week of the disease?

L. Agglutination-lysis titre
M. Agglutination titre
N. Blood culture plus animal inoculation
O. Darkground microscopy of the blood
P. Urine culture.

Answers overleaf

4.35 B. The Frei test detects delayed hypersensitivity type IV skin reaction to intradermal group A chlamydial antigen. Conversion to positive develops by 6 weeks and remains positive for life. Complement fixation tests and newer tissue culture methods for isolation of group A chlamydiae are also available for laboratory diagnosis.

4.36 G. Patients with infectious mononucleosis (glandular fever) develop characteristic heterophile antibodies which agglutinate sheep erythrocytes, and detection of these antibodies forms the basis of the Paul-Bunnell reaction. A rapid slide screening test is now available.
Normal individuals and persons with serum sickness following injections of horse serum may produce similar heterophile antibodies unrelated to infectious mononucleosis, but these can be recognized by differential absorption of the patient's serum with ox cells and with guinea pig kidney. Infectious mononucleosis antibodies are absorbed by ox cells, normal heterophile antibodies are absorbed by guinea pig kidney and serum sickness antibodies are absorbed by both.

4.37 N. Leptospira are fairly numerous in the blood early in the disease but rapidly disappear after the first week, although they may be demonstrated in the urine by darkfield microscopy, culture on serum broth, and by animal inoculation. Direct examination of the blood is unreliable due to difficulty in differentiating fibrin strands and erythrocyte disintegration products. Antibody tests become positive by the 10th day and rising titres may be demonstrated by agglutination of killed standard suspensions or by agglutination and lysis of living leptospira.

4.38 Which diagnostic test for brucellosis in man is of LEAST practical value and therefore rarely used?

Q. Blood culture
R. Brucellin skin test
S. Complement fixation test
T. Direct and indirect agglutination tests
U. Marrow culture.

4.39 Which technique is routinely used for the serological diagnosis of enteric fever and brucellosis but NEVER for syphilis?

V. Complement fixation test
W. Counter-current immunoelectrophoresis
X. Precipitin reaction in agar-gel
Y. Slide agglutination reaction
Z. Tube agglutination reaction.

4.40 Which genus of *Enterobacteriaceae* includes strains which are specifically agglutinated by serum of patients with typhus fever?

A. *Escherichia*
B. *Proteus*
C. *Salmonella*
D. *Serratia*
E. *Shigella.*

4.41 For which purpose are paired acute and convalescent serum samples taken?

F. Assay serum antibiotic levels
G. Detect fourfold increase in specific antibody titre
H. Determine whether recovery is complete
J. Exclude development of chronic infective state
K. Show antibody titre has fallen with recovery.

Answers overleaf

4.38 R. Brucellin tests resemble the tuberculin reaction in demonstrating delayed hypersensitivity without indicating active disease. Up to 30 per cent of the normal population give a positive reaction and negative reactions occur even in the presence of culturally proven active disease, so that the test is unreliable.
Cultivation of the organism from blood or bone marrow is completely reliable where positive, but may prove extremely difficult in chronic disease. Agglutination reactions are routinely done in all suspected cases. Tests used are direct agglutination, the indirect or anti-human globulin (AHG or Coomb's) test and complement fixation. Active chronic infection usually shows as a high AHG titre which detects IgG whereas acute cases usually have a raised direct agglutination which reflects an IgM response.

4.39 Z. The Widal reaction for serological diagnosis of enteric fever is based upon detection of specific "H", "O" and where appropriate "Vi", agglutinins, against *S. typhi* and *S. paratyphi*, A, B and C by tube agglutination. Originally, the Widal reaction was introduced as a tube agglutination test for *S. paratyphi B.*

4.40 B. Certain strains of *P. vulgaris* share somatic antigens with *R. prowazeki* (epidemic typhus), *R. mooseri* (endemic murine typhus) and *R. rickettsi* (Rockey mountain spotted fever) and are agglutinated by sera from infected individuals. Proteus strains OX19, OX2 and OXK have been used as test antigens for the serological diagnosis of the typhus fevers. Brill-Zinsser recrudescent typhus fever invariably is negative.

4.41 G. Evidence of a specific response to an infecting organism is of diagnostic value where the antibody is not normally present in the population and where the preferred isolation of the causal agent may prove difficult or unsuccessful. A single test is of limited value as it may represent antibody from an earlier and now resolved infection, response to prior immunization or an anamnestic response to cross-reacting antigens. More significance can be placed on demonstration of greater than fourfold increase in antibody titre between acute-phase serum, taken at the beginning of the illness and the convalescent serum, taken 2—3 weeks later.

5. Antimicrobial Agents

5.1 **Which statement about para-aminobenzoic acid (PABA) is FALSE?**

A. Acts as growth factor
B. Deficiency is rapidly bactericidal
C. Precursor of folic acid
D. Sulphonamides compete with PABA
E. Yeast extract is rich in PABA.

5.2 **Which one of the following antimicrobial agents is neutralized by para-aminobenzoic acid?**

F. Formaldehyde
G. Nalidixic acid
H. Nitrofurantoin
J. Sulphamethoxazole
K. Trimethoprim.

5.3 **Which one is *NOT* bactericidal in action?**

L. Amoxicillin
M. Cephaloridine
N. Isoniazid (INH)
O. Para-amino salicylic acid (PAS)
P. Tobramycin.

Answers overleaf

5.1 B. Deficiency of PABA interferes with the formation of the vitamin folic acid which has only a bacteriostatic effect on the organisms. The structural similarity of sulphonamides and PABA explains why they compete with PABA and inhibit folic acid synthesis.

5.2 J. Sulphonamides interfere with folate metabolism by substrate competition. Excess of PABA neutralizes the effect as does the end product, folic acid.
Trimethoprim acts at a later stage in folic acid synthesis and is unaffected by PABA although reversal of inhibition can be produced by thymidine.
The combination of sulphamethoxazole and trimethoprim in co-trimoxazole produces a double blockade on the folic acid biosynthetic pathway which limits emergence of resistance, gives a synergistic effect and converts the individual bacteriostatic activities to combined bactericidal.

5.3 P. PAS, like the sulphonamides, is a structural analogue of PABA. Its bacteriostatic action is due to competitive inhibition of PABA — folate metabolism.
Amoxicillin and cephaloridine inhibit cross-linking between wall units with ultimate rupture and cell death. Isoniazid is restricted in use to combined treatment of tuberculosis. It prevents synthesis of mycolic acid which results in a structurally weak wall. This produces a bactericidal effect except in low concentration when it is bacteriostatic. Tobramycin is an aminoglycoside related to gentamicin with a similar effect due to irreversible blockade of protein synthesis at the ribosomal level.

5.4 Which chemotherapeutic agent strikingly potentiates the antibacterial activity of sulphonamides?

Q. Isoniazid
R. Metronidazole
S. Nalidixic acid
T. Nitrofurantoin
U. Trimethoprim.

5.5 Which form of penicillin is acid-labile and inactivated by penicillinase?

V. Ampicillin
W. Cloxacillin
X. Methicillin
Y. Penicillin G
Z. Penicillin V.

5.6 What is the particular value of probenecid in antimicrobial therapy?

A. To antagonize penicillinase activity
B. To exert a synergistic action with penicillin
C. To increase penicillin's antibacterial spectrum
D. To maintain higher blood levels of penicillin
E. To reduce protein binding of penicillin.

5.7 In which phase of growth is a cephalosporin MOST effective in its bactericidal action?

F. Decline
G. Lag
H. Log
J. Stationary
K. Uniform throughout growth cycle.

Answers overleaf

5.4 U. Sulphonamides and trimethoprim act sequentially to block the biosynthetic pathway of the vitamin folic acid. Sulphonamides block incorporation of PABA into dihydrofolate. Trimethoprim inhibits dihydrofolate reductase thus preventing conversion of dihydrofolate into tetrahydrofolate, the functional form of folic acid. This sequential blockade gives a synergistic effect in terms of potency, increases spectrum of activity, converts to bactericidal action and reduces the possibility of development of resistance to either agent.

5.5 Y. Penicillin G (benzyl-) is acid-labile and its beta-lactam ring is split by penicillinase. Ampicillin and penicillin V (phenoxymethyl-) are both sensitive to penicillinase but are relatively acid stable (that is why they can be administered orally). Cloxacillin and methicillin are resistant to the action of penicillinase with cloxacillin acid stable and the injection form, methicillin, acid labile.

5.6 D. During treatment with benzyl penicillin, 60 per cent of the dose is lost by renal tubular excretion. While this may be acceptable for infections caused by highly sensitive organisms in accessible tissue sites, it is unsatisfactory with organisms of only moderate sensitivity or in serious infections affecting sites into which antibiotics penetrate poorly. For these it is essential that a constantly high level of penicillin is maintained and this may be achieved most easily by the addition of probenecid which blocks the wasteful tubular excretion of penicillin. This combination has been widely advocated for treatment of *Strep. viridans* subacute bacterial endocarditis.

5.7 H. Cephalosporins act like penicillin by preventing the cross-linkage of mucopeptide strands in the growing cell wall and is therefore most active during exponential growth.

5.8 **Which does *NOT* act by inhibition of cell-wall synthesis?**

L. Bacitracin
M. Cephaloridine
N. Cycloserine
O. Tetracycline
P. Vancomycin.

5.9 **Which is bactericidal in action?**

Q. Chloramphenicol
R. Erythromycin
S. Gentamicin
T. Lincomycin
U. Tetracycline.

5.10 **Which organism, from the clinical standpoint, is NEVER resistant to penicillin?**

V. *Haemophilus influenzae*
W. *Mycoplasma pneumoniae*
X. *Pseudomonas aeruginosa*
Y. *Staphylococcus albus*
Z. *Streptococcus pyogenes.*

Answers overleaf

5.8 O. Tetracyclines inhibit protein synthesis by binding to both ribosomes and mRNA. This prevents the binding of aminoacyl — tRNA to the mRNA — ribsomes complex which blocks protein synthesis, producing a bacteriostatic effect. The other antibiotics inhibit all three stages of cell wall biosynthesis: *cycloserine* inhibits stage I, acting within the cell interior to inhibit precursor formation; *bacitracin* and *vancomycin* inhibit stage II, bound to the cell membrane where they block the transport of the basic repeating unit through the membrane where it would normally be incorporated into the cell wall; *cephaloridine*, like penicillin, inhibits stage III at the cell exterior where it prevents cross-linking of the linear mucopeptide polymer.

5.9 S. Gentamicin irreversibly blocks bacterial protein synthesis by binding to the 30S subunit of the bacterial 70S ribosomes.
The other antibiotics are merely bacteriostatic. Success in treatment with them requires a normal phagocytic mechanism to complete the killing action.

5.10 Z. Despite the extensive clinical use of penicillin since its introduction, *Strep. pyogenes* still shows no tendency to develop resistance and penicillin remains the drug of choice in infections caused by this organism.
H. influenzae shows only moderate sensitivity to penicillin *in vitro* and infections prove unresponsive, although highly sensitive to ampicillin. *Myc. pneumoniae* lacks the sensitive rigid cell wall mucopeptide and is completely resistant.
Ps. aeruginosa is resistant to benzyl penicillin but may be highly sensitive to carbenicillin, although resistant strains, often with transmissible R-factors, are increasingly found in hospital practice. *Staph. albus* quite often is found to be resistant, either due to production of penicillinase or to some inherent mechanism.

5.11 Which of the following antimicrobial agents is LEAST likely to be effective in post-operative wound sepsis caused by a "hospital strain" of *Staphylococcus aureus?*

A. Benzyl-penicillin
B. Erythromycin
C. Flucloxacillin
D. Fucidin
E. Lincomycin.

5.12 Which example of antimicrobial therapy prescribed for an initially sensitive strain, is MOST likely to promote emergence of resistant mutants with resultant clinical relapse?

F. Cloxacillin for *Staphylococcus aureus* breast abscess
G. Metronidazole for *Bacteroides fragilis* wound infection
H. Penicillin G for *Staphylococcus aureus* septicaemia
J. Streptomycin for chronic pulmonary tuberculosis
K. Tetracycline for *Mycoplasma pneumoniae* primary atypical pneumonia.

5.13 Which organism may acquire a high degree of drug-resistance by one step mutation, with associated loss of catalase activity, on exposure to a certain chemotherapeutic agent?

L. *Mycobacterium tuberculosis*
M. *Pseudomonas aeruginosa*
N. *Shigella dysenteriae* (F-strains)
O. *Shigella flexneri* (F+strains)
P. *Staphylococcus aureus* (hospital strains).

5.14 Which one is resistant to metronidazole (Flagyl)?

Q. *Bacteroides fragilis*
R. *Candida albicans*
S. *Entamoeba histolytica*
T. *Trichomonas vaginalis*
U. *Vincent's organisms.*

Answers overleaf

5.11 A. Hospital acquired infection with *Staph. aureus* is most frequently caused by penicillinase-producing strains which predominate in the hospital environment due to the selective pressure exerted by the use of penicillin. Therefore, benzyl-penicillin is unlikely to be effective and one of the penicillinase-resistant penicillins would be the drug of choice unless the patient was hypersensitive when any of the other drugs could be used.
Had the strain proved on testing to be penicillin-sensitive then benzyl-penicillin would have been indicated as it is much more active than the modified penicillinase-resistant forms.

5.12 J. *M. tuberculosis* rapidly produces streptomycin-resistant mutants of both high and low resistance. For streptomycin to be effective it must successfully enter the cell and bind to its site of action on the ribosome. Low level resistance results from modification of the cell membrane preventing uptake of the drug. High level resistance results from a single-step mutation which alters the ribosomal protein thus preventing essential binding of streptomycin to the ribosome. Treatment against tuberculosis should always use a combination of drugs concomitantly to prevent, or at least delay, the emergence of resistant mutants.

5.13 L. When isoniazid is used alone for treatment of tuberculosis, catalase positive (low resistance) and catalase negative (high resistance) mutants regularly develop by single-step mutation. Despite a longer generation time and increased susceptibility to H_2O_2 within phagocytes, they are capable of establishing infection and of producing progressive tuberculosis in man.

5.14 R. *Candida albicans* is resistant to antibacterial antibiotics and chemotherapeutic agents but is sensitive to nystatin, amphotericin B and 5-fluorocytosine.
Metronidazole is highly active against anaerobes and is also effective in amoebiasis and trichomoniasis. It is the drug of choice against *B. fragilis* and may be used for Vincent's infection in patients who are allergic to penicillin.

5.15 In which infection is it highly desirable that the minimum inhibitory concentration (MIC) and the minimum bactericidal concentration (MBC) are determined for the antibiotic of choice against the clinical isolate?

V. Scarlet fever
W. Subacute bacterial endocarditis
X. Syphilis
Y. Tetanus
Z. Upper respiratory tract infection.

5.16 Which one of the following is associated with an increased incidence of gentamicin ototoxicity?

A. Acute uncomplicated pyelonephritis
B. Combined therapy with gentamicin and cephaloridine
C. Concurrent treatment with diuretics
D. Pre-existing renal impairment
E. Unilateral nephrectomy for renal carcinoma.

5.17 Which one of the following is *NOT* an adverse side effect of prolonged antimicrobial therapy?

F. Emergence of resistant strains
G. Diarrhoea
H. Drug addiction
J. Pulmonary eosinophilia
K. Uraemia.

Answers overleaf

5.15 W. MIC and MBC are tube dilution sensitivity tests used to determine the least amount of antibiotic which respectively inhibits or kills the test organism. In SBE it is essential to prescribe a bactericidal antibiotic to kill and not simply inhibit the pathogen. By determining the MBC a suitable dosage can be calculated to obtain serum levels in excess of the minimum bactericidal concentration.

5.16 D. Irreversible labyrinthine intoxication only occurs when serum gentamicin levels exceed 10—12 mcg/ml which is the safe upper limit. A single dose of 80 mg gentamicin gives the same peak levels in normal individuals and in patients with renal impairment. With renal impairment, renal clearance is slow and 80 mg given at the normal interval would produce levels in excess of the safe upper limit. By retaining the same unit dose of 80 mg but extending the interval from 8 to 12 hours, if there is moderate impairment, or 48 hours with severe impairment, therapeutic levels may be maintained without toxicity. Serum levels should be regularly monitored and the dose interval (or even the dose) modified accordingly.
Combined therapy increases the risk of nephrotoxicity only if there is also renal impairment or doses excessive.

5.17 H. Antibiotics and chemotherapeutic agents are not drugs of addiction although patients may become dependent upon them for long term prophylaxis to limit the frequency and severity of exacerbations of chronic infection. This is true of patients with progressive obstructive airway disease or after bile duct or ureteric bypass operations to allow direct excretion into the gut. In such patients, local and systemic infections frequently recur or relapse after a period of remission and the value of prophylactic measures is debatable.

5.18 Which antibiotic has been MOST frequently implicated in the production of pseudomembranous enterocolitis?

L. Ampicillin
M. Chloramphenicol
N. Clindamycin
O. Gentamicin
P. Tetracyclines.

5.19 For which of the following is chloramphenicol still probably the drug of choice?

Q. Bacillary dysentery
R. Carrier state of typhoid fever
S. Salmonella food poisoning
T. Staphylococcal food poisoning
U. Typhoid fever.

5.20 For which clinical infection is antibiotic therapy usually *NOT* recommended?

V. Acute cholecystitis
W. Gastroenteritis caused by *Salmonella typhimurium*
X. Peritonitis secondary to appendicectomy
Y. Staphylococcal enterocolitis
Z. Typhoid fever.

Answers overleaf

5.18 N. Most incidents of antibiotic-associated pseudomembranous enterocolitis have occurred during treatment with lincomycin or clindamycin (7-chloro-lincomycin) although it has occurred rarely with chloramphenicol and ampicillin. Symptoms include fever, diarrhoea and cramping abdominal pain. The intestine shows extensive epithelial necrosis and if the drug is not discontinued, the condition could prove fatal. The pathogenesis is unknown but the epidemiological pattern of associated individuals in the same family or the same hospital ward suggests a pharmacogenetic mechanism linked with some infective agent.

5.19 U. For more than 20 years chloramphenicol has been the drug of choice in typhoid fever and other severe salmonella infections. It is effective both in the early and late stages of typhoid fever, rapidly lowering the fever to normal but promotes the tendency to relapse and may possibly encourage chronic carriage. The occasional serious toxic effect on the marrow of neutropenia and fatal aplasia has stimulated the search for an effective alternative of which co-trimoxazole, ampicillin and recently mecillinam seem the best. Chloramphenicol is contra-indicated for any of the other diseases listed, including chronic typhoid carriers for which ampicillin is recommended, producing 80 per cent cure rate.

5.20 W. "Food poisoning" gastroenteritis caused by non-invasive salmonellae is usually relatively mild and self-limiting. Exhibition of antibiotics may prolong rather than shorten the period of carriage. Antibiotic therapy would be indicated in serious infection, and in debilitated or very young subjects.

5.21 Which one of the following chemotherapeutic agents is a major cause of the rare, but serious, Stevens-Johnson syndrome?

A. Co-trimoxazole
B. Erythromycin
C. 5-Fluorocytosine
D. Fusidic acid
E. Long-acting sulphonamides.

Answers overleaf

5.21 E. The Stevens-Johnson syndrome is a very rare (incidence of 1 in 10 million doses) but serious immunologically-mediated toxic reaction to sulphonamides, although it may also result from other causes. Preparations particularly incriminated are long-acting sulphonamides which remain in the circulation for long periods, bound to plasma proteins when they act as haptens. The syndrome is characterized by a florid stomatitis with extensive ulceration of the lips and skin rashes with exfoliation, and is described as a potentially fatal form of erythema multiforme. Withdrawal of the inciting drug is mandatory but if a long-acting preparation has been used the toxic reaction may even develop after stopping treatment, due to persistence of sulphonamide in the blood stream.

6. Prophylactic Immunization

6.1 In which disease would you be MOST surprised to find active immunity in a patient who had never been artificially immunized and had never had a clinically apparent attack of the disease?

A. Diphtheria
B. Primary tuberculosis
C. Scarlet fever
D. Tetanus
E. Whooping cough.

6.2 A clinician suspects an immigrant child has diphtheria. What should be wait for before initiating treatment with antitoxin?

F. Bacteriological identification of *Corynebacterium diphtheriae*
G. Demonstration of toxigenicity by the clinical isolate
H. First signs of cranial nerve palsies
J. Formation of typical pseudomembrane
K. Result of skin and ophthalmic tests for serum hypersensitivity.

6.3 Which item of information clearly demonstrates that an individual lacks protective immunity against diphtheria?

L. Avirulent *Corynebacterium diphtheriae* isolated from a throat swab
M. Gave a positive Schick reaction
N. No history of previous immunization with toxoid
O. Reliable history of immunization with "triple vaccine"
P. Virulent *Corynebacterium diphtheriae* isolated from a throat swab.

Answers overleaf

6.1 D. Surprisingly, active immunization against tetanus rarely follows recovery from clinical tetanus, and subclinical or latent infection capable of conferring immunity does not occur. Recovery from the disease must therefore be followed by a full course of artificial active immunization with tetanus toxoid.
The other diseases, in either clinical or subclinical form, regularly confer active immunity due to production of neutralizing antitoxin (diphtheria and scarlet fever), antibacterial antibody (whooping cough) or cell-mediated immunity (tuberculosis) following induction of a state of delayed hypersensitivity to tuberculo-proteins.

6.2 K. Any patient suspected of having diphtheria must be given antitoxin unless the skin and ophthalmic tests show serum hypersensitivity (tests take about 30 minutes) in which case they should first be desensitized. Prognosis is directly proportional to the day of the disease when antitoxin is first administered and a delay of 24 or 48 hours, while waiting for the result of nasal and throat swab culture, may mean the loss of life.

6.3 M. A positive reaction to an intradermal injection of diphtheria toxin indicates that little or no circulating antibody is present and that the subject is susceptible to diphtheria toxin and therefore the disease diphtheria. Carriers of *Corynebacterium diphtheriae* give a negative reaction whereas early cases of diphtheria react positively.

6.4 What does a positive Schick test indicate?

Q. Confirms a clinical attack of diphtheria
R. Excludes the possibility of diphtheria
S. Indicates the individual is a carrier
T. Reveals susceptibility to diphtheria
U. Shows that immunization is not required.

6.5 Which one of the following is used to confer active immunity to diphtheria?

V. Diphtheria antitoxin
W. Diphtheria toxin
X. Diphtheria toxoid
Y. Live, attenuated diphtheria bacilli
Z. Killed, virulent diphtheria bacilli.

6.6 In which situation is the administration of TAB vaccine contraindicated?

A. Home contacts of chronic typhoid carrier
B. Individuals in recent contact with a case of typhoid fever
C. Laboratory worker in a salmonella reference laboratory
D. Military personnel posted to an endemic typhoid area
E. Nursing staff of an infectious diseases hospital.

Answers overleaf

6.4 T. Susceptibility to diphtheria is indicated by the appearance at 48 hours of a 1—5 cm inflamed and indurated area at the site of the intradermal injection of a standard skin-test-dose of diphtheria toxin provided the site injected with heat-inactivated (negative-control) toxin shows no reaction. A positive Schick test indicates the individual has insufficient circulating antitoxin to neutralize diphtheria toxin and requires to be immunized. Schick tests and throat cultures are usually done on contacts. Positive culture and positive Schick test indicates potential cases in need of treatment. Positive culture with negative Schick test indicates the carrier state. Negative culture with a positive Schick reveals susceptibility to diphtheria. Negative culture with a negative Schick test shows the individual to be safe and not a public health risk.

6.5 X. Immunity to diphtheria is dependent upon possession of an adequate level of circulating antitoxin. Artificial active immunity is acquired by the administration of a suitable course of diphtheria toxoid. This is currently offered to children as the "triple vaccine" (diphtheria, tetanus, pertussis) in the first year of life, beginning at 3 months, and continued into childhood, reinforced by boosters. The Schick test may be used to determine if immunization has been effective in raising the level of circulating antitoxin above the threshold of 0.03 units/ml which determines whether the individual is Schick negative and immune to diphtheria.

6.6 B. Contacts are kept under surveillance during the 3 week incubation period (usually 10—14 days with a range from 5—21 days) before being cleared by negative stool culture. Contacts who become ill or develop a fever would be examined intensively by blood, faeces and urine culture as well as serology by Widal reaction. Immunization given during the incubation period would nullify any serological changes even if it did not temporarily compromise resistance as suggested by earlier workers (provocation typhoid).

6.7 In which disease does the level of specific agglutinin, opsonin and bactericidal antibody correlate badly with the level of immunity to infection?

F. *Haemophilus influenzae* meningitis
G. Meningococcal meningitis
H. Pneumococcal lobar pneumonia
J. Typhoid fever
K. Whooping cough.

6.8 Which vaccine is comprised of dead organisms?

L. BCG
M. Cholera
N. Sabin poliomyelitis
O. Smallpox
P. Yellow fever strain 17D.

6.9 Against which infectious disease has the use of killed bacterial vaccines afforded no significant protection?

Q. Cholera
R. Epidemic typhus fever
S. Shigellosis
T. Typhoid fever
U. Whooping cough.

6.10 Which NEVER results as a complication of prophylactic active immunization?

V. Encephalitis
W. Local abscess formation
X. Provocation poliomyelitis
Y. Serum sickness
Z. Systemic infection with vaccine.

Answers overleaf

6.7 J. Individuals with a good response to TAB vaccine may still contract typhoid fever although generally less severe. Even after typhoid fever with high antibody titres a second infection may occur, and chronic infection in bone, gall-bladder and other sites may persist. The other diseases are unlikely to occur in persons with specific antibody to capsular and surface antigens, and in established infection resolution usually follows appearance of protective antibody (e.g. resolution by crisis in pneumococcal lobar pneumonia).

6.8 M. Killed cholera vaccine gives limited protection from about ten days after inoculation. The vaccine protects against both classical *V. cholerae* and the El Tor biotype. Attempts to develop a more effective *toxoid* vaccine promise to give a more effective prophylactic than the whole cell vaccine.

6.9 S. Bacillary dysentery is neither prevented nor modified in severity by immunization with killed shigella vaccine probably due to lack of stimulation of specific IgA antibody. In the other diseases the use of killed vaccine has afforded protection sufficient to prevent or modify the disease.

6.10 Y. Serum sickness is caused by administration of a large amount of foreign protein or of a drug (e.g. penicillin) that can bind to the patient's proteins altering their immunological specificity. It was first described in treatment of diphtheria with antitoxin of horse origin, hence serum sickness. Following the injection symptoms appear 1–2 weeks later when circulating soluble antigen — antibody complexes cause widespread Arthus reactions, with fever, urticaria, swollen painful joints, albuminuria, and even frank glomerulonephritis.
The other complications occasionally follow active immunization.

6.11 Which vaccine should *NOT* be given to individuals on immunosuppressive therapy?

A. Cholera
B. Salk poliomyelitis
C. Smallpox
D. Typhoid-paratyphoid (TAB)
E. Whooping cough.

Answers overleaf

6.11 C. Generalized vaccinia may follow administration of the live smallpox vaccine to individuals with suppressed cell-mediated immunity, either from primary immunological deficiency or secondary to disease which affects the lymphoid tissues (e.g. leukemia or lymphoma) or as a result of immunosuppressive therapy. After vaccination, the local primary lesion fails to heal and progressive local extension finally results in systemic dissemination of vaccinia virus with usually a fatal outcome. (Methisazone therapy may be of some value).

Another example of progressive lesions following the use of a living vaccine in persons with suppressed immunity is the fatal giant-cell pneumonia caused by attenuated measles virus. These examples emphasize that persons with congenital or acquired immunological deficiency must not receive live vaccines.

Part Four

OTHER FORMS OF MULTIPLE CHOICE QUESTIONS

Two responses are required to each of the following sets of questions. From the numbered list, select and identify by number the item which is specifically associated with 4 of the 5 items in the alphabetical list and identify the unrelated item by letter.

1.1 Lobar pneumonia
1.2 Plasma cell pneumonia
1.3 Primary atypical pneumonia.

A. Adenovirus
B. *Chlamydia psittici*
C. *Coxiella burneti*
D. *Mycoplasma pneumoniae*
E. *Pneumocystis carinii*

2.1 Lymphocyte
2.2 Macrophage
2.3 Neutrophil polymorph.

F. Adheres to immune complexes
G. Capable of activation by lymphokines
H. Capable of mitosis and division
J. Has receptors for C3
K. Releases chemotactic factors.

3.1 Decreased humoral response
3.2 Impaired lymphocyte transformation
3.3 Lowered serum haemolytic complement.

L. Active syphilis
M. Di George's syndrome
N. Exacerbations of rheumatic heart disease
O. Measles
P. Tuberculoid leprosy.

Answers overleaf

1. (1.3 — E) Primary atypical pneumonia is a syndrome of multiple aetiology. *Myp. pneumoniae* (Eaton's agent) is usually responsible but other causes include those listed plus influenza, para-influenza and respiratory syncytial viruses. *Myp. pneumoniae* alone stimulates production of cold agglutinin and antibody to streptococcus MG. *Pneumocystis carinii* is an opportunistic protozoon which causes plasma-cell pneumonia in babies and persons with compromised defences.

2. (2.2 — K) Chemotactic factors for macrophages are released by both neutrophils and lymphocytes but macrophages do not themselves produce such factors. Neutrophils are capable neither of mitosis nor of activation by lymphokines. Lymphocytes show all five properties.

3. (3.2 — P) Assay of lymphocyte transformation is done when investigating possible immunological deficiency or monitoring immunosuppression. In evaluating such tests it must be remembered that in addition to congenital immunological deficiency many acquired conditions also impair the reactivity of thymic derived lymphocytes (e.g. aging, malnutrition, uraemia, major surgery, marihuana smoking and lymphoid tumours, as well as a variety of virus and bacterial infections which involve the lymphoid organs).

4.1 Cestode
4.2 Nematode
4.3 Trematode.

Q. *Ancylostoma duodenale*
R. *Ascaris lumbricoides*
S. *Echinococcus granulosus*
T. *Enterobius vermicularis*
U. *Trichuris trichiura.*

5.1 *Echinococcus granulosus*
5.2 *Entamoeba histolytica*
5.3 *Fasciola hepatica.*

V. Associated with liver abscess
W. Diagnostic serological tests available
X. Fresh water snails are intermediate host
Y. Mature thin-walled cysts contain four nuclei and chromatoid bodies
Z. Vegetative forms may contain erythrocytes.

6.1 Cell membrane
6.2 Cell wall
6.3 Spore cortex.

A. Actively transports ions, sugars and amino acids
B. Oxygen tension affects its composition
C. Participates in DNA replication
D. Site of electron transport system
E. Supporting structure with no osmotic function.

7.1 *Francisella tularensis*
7.2 *Haemophilus influenzae*
7.3 *Yersinia pseudotuberculosis.*

F. Common in Great Britain
G. Gram-negative nonmotile coccobacillus
H. Produces granulomas resembling tubercles
J. Requires blood enriched media
K. Transmitted by contact with infected animals.

Answers overleaf

4. (4.2 — U) *E. granulosus* (tapeworm) is a cestode, causing hydatid disease in man infected by the cystic stage. The others are nematodes (roundworms).
   A. duodenale (hookworm) occurs mainly in Asian immigrants, sometimes causing anaemia. *A. lumbricoides* (roundworm) is not uncommon in the U.K. Usually harmless, it occasionally blocks the biliary and pancreatic ducts, or pierces the intestinal wall. *E. vermicularis* (pinworm or threadworm) usually occurs in children and is perpetuated by autoinfection from anal scratching. *T. trichuria* (whipworm) rarely causes symptoms. The barrel-shaped cysts with terminal plugs are very easily identified.

5. (5.2 — W) Man is the natural host and infection is normally transmitted by contamination of food and water with faeces containing viable cysts of *E. histolytica.*
   The "odd man out" refers to *F. hepatica* whose primary host is the sheep and intermediary host, the fresh water snail. Man is only rarely infected.

6. (6.1 — E) The cell membrane maintains the cell's internal environment by active transport of ionized and small non-ionized molecules, thus functioning as an osmotic barrier. It has an active role in growth and division, transporting and modifying preformed cell wall precursors and acting as an attachment site for replicating DNA. Ingrowth of the membrane separates the two nuclei and effects septation. Mechanical support is provided by the outer rigid framework of cell wall mucopeptide.

7. (7.1 — F) *F. tularensis* does not occur naturally in Britain. *H. influenzae* elicits an acute inflammatory response without granuloma formation and is transmitted exclusively by man. *Y. pseudotuberculosis* is flagellate and motile when cultured at 22°C (room temperature) and grows well in the absence of blood products (e.g. on MacConkey's medium).

8.1	Food-borne disease	L.	Food poisoning
8.2	Notifiable disease	M.	Infective jaundice
8.3	Zoonosis.	N.	Mumps
		O.	Tuberculosis
		P.	Whooping cough.

Mark the following statements in accordance with the key below:

A. If assertion and reason are true and cause and effect related
B. If assertion and reason are true but *NOT* cause and effect related
C. If the assertion is true but the reason is false
D. If the assertion is false but the reason is true
E. If both the assertion and reason are false.

9. A negative slide coagulase reaction does not exclude *Staphylococcus aureus*
because
coagulase production in some strains can be detected only by a tube coagulase test.

10. *Cryptococcus neoformans* (*Torula histolytica*) is known to produce a powerful local lytic effect on infected tissues
because
a prominent wide clear zone surrounds the budding yeast cells in cryptococcosis.

11. S $\rightarrow$ R mutation is usually associated with loss of mucoid colonies and formation of rough colonies
because
rough mutants show a distinct tendency to autoagglutinate.

Answers overleaf

8. (8.2 — N) The Health Services and Public Health Act, 1968, requires medical practitioners to make formal notification of any of the listed communicable diseases to the appropriate Health Authority.

9. A. *Staph. aureus* should not be excluded by a negative slide test as only a proportion of strains possess the necessary surface "clumping factor". Negative reactors should be tested by the slower but more reliable tube method, before discarding as *Staph. albus.*

10. D. The characteristic clear zone is due to formation of a prominent polysaccharide capsule which is also seen with cells in spinal fluid and in culture. The zone was originally misinterpreted as representing local tissue liquefaction, hence the old name *T. histolytica.*

11. B. Smooth mucoid colonies are produced only by capsulated organisms. Bacterial variation associated with loss of the ability to form capsules results in the formation of dry, irregular colonies.

12. **Farmer's lung is an opportunistic fungal infection occurring mainly in association with agammaglobulinaemia**
because
Micropolyspora faeni **rapidly invades the lungs of persons lacking protective serum IgG and secretory IgA.**

13. **Immunization and chemoprophylaxis may protect man from plague but cannot eradicate the disease**
because
chronic human carriers perpetuate the disease in the human population.

Answer the questions according to the following key:

A. If only 1, 2 and 3 are correct
B. If only 1 and 3 are correct
C. If only 2 and 4 are correct
D. If all are correct
E. If none is correct.

14. **Feature of schistosoma**

    1. Final stage larva from the fresh water snail is a cercaria
    2. First stage larva from the hatched egg is a miracidium
    3. Man harbours the adult worms which liberate eggs
    4. Man is infected by ingestion of encysted cercaria contaminating water plants, fish and molluscs.

Answers overleaf

12. E. Farmer's lung is *NOT* an infection but an allergic alveolitis to inhaled spores from fungi present in mouldy hay. Formation of antigen-antibody complexes with precipitating IgG produces a type III hypersensitivity reaction in the lung. This could not happen in the absence of specific antibody.
Opportunistic fungal infections with *M. faeni* do not occur and those caused by other fungi are associated with deficiencies in cell-mediated immunity.

13. C. Sylvatic plague can be eradicated only by the complete extermination of susceptible wild rat and other rodent populations which are the natural reservoir of infection. As most human epidemics arise from urban plague in the domestic rat population, control measures have been largely directed at killing rat fleas by insecticide, rat extermination by Warfarin and protection of humans by immunization with killed or attenuated *Yersinia pestis* vaccines, supplemented by chemoprophylaxis.

14. A. Schistosoma differ from other trematodes by having no encysted metacercarial stage.
Man is infected by contact with water containing cercaria which penetrate the skin and migrate to the portal vein. Flukes mature in the intrahepatic portal vein and migrate to the pelvic venous plexuses where they copulate and liberate eggs in either the urine (*S. haematobium*) or faeces (*S. mansoni* or *S. japonicum*). Eggs hatched in water release a free-swimming miracidium which infects the intermediate host, the fresh water snail. Following multiplication as sporocysts, the final infective larval stage (cercaria) is liberated to complete the life cycle.

15. **Associated with schistosomiasis (bilharziasis)**

    1. Fever, urticaria and eosinophilia
    2. Local granulomas
    3. Painless terminal haematuria
    4. Vesical carcinoma.

16. **Aetiological agent of protozoal infection**

    1. *Leishmania donovani* (kala-azar)
    2. *Trypanosome cruzi* (Chagas' disease)
    3. *Trypanosome rhodesiense* (sleeping sickness)
    4. *Wuchereria (Filaria) bancrofti* (filariasis)

17. **Interval between paroxysms of fever in malaria**

    1. 3 days in benign tertian malaria (*P. vivax*)
    2. 3 days in malignant tertian malaria (*P. falciparum*)
    3. 3 days in ovale tertian malaria (*P. ovale*)
    4. 4 days in quartan malaria (*P. malariae*)

Answers overleaf

15. D. Toxic products from the growing worms elicit a systemic allergic response and local reaction to the eggs frequently produces granulomas and later fibrosis. The sharp spined eggs are prone to cause bleeding and chronic irritation of the bladder may induce malignant change.

16. A. Even if you were not familiar with the specific names of these parasites you should have recognized that filariasis is not due to a protozoon but results from infestation with a nematode. The adult worms inhabit man's lymphatics and may produce lymphatic blockage with the syndrome of elephantiasis.

17. E. Paroxysms of fever in malaria result from cycles of parasitic multiplication within erythrocytes (erythrocytic schizogony). The terms tertian and quartan represent the periodicity of the febrile paroxysms counting both days of fever and the interval between. Thus, periodic fever recurs every 2 days in tertian and every 3 days in quartan malaria.